AF371498

Engineering Materials

For further volumes:
http://www.springer.com/series/4288

Hamid Bentarzi

Transport in Metal-Oxide-Semiconductor Structures

Mobile Ions Effects on the Oxide Properties

 Springer

Hamid Bentarzi
Dept. of Electrical and Electronic Engineering
University of Boumerdes
Freedom street
35000 Boumerdes
Algeria
e-mail: bentarzi_hamid@yahoo.com

ISSN 1612-1317   e-ISSN 1868-1212

ISBN 978-3-642-16303-6   e-ISBN 978-3-642-16304-3

DOI 10.1007/978-3-642-16304-3

Springer Heidelberg Dordrecht London New York

Cover design: deblik, Berlin

Printed on acid-free paper

Springer is part of Springer Science+Business Media (www.springer.com)

Preface

The present work deals with the study of the mobile ions in the silicon dioxide insulator, which has great importance because their presences affect significantly on the MOS structure characteristic. The subject is introduced with the necessary background concepts of MOS structure dealing with various aspects of the oxides and their charges. Besides, theoretical approaches to determine the density of mobile ions as well as their density-distribution along the oxide thickness are developed. In fact, three attempts have been discussed each makes use of different approaches. In the first attempt, the density of the mobile ions has been determined from experimental measurements using different techniques such as the Charge Pumping (CP) technique associated with the Bias Thermal Stress (BTS) method. In the second attempt, the theoretical approaches using empirical models or numerical approach for the mobile ions density distribution are described. In the last attempt, an analytical model of the mobile ions density distribution, which is based on physical concepts at equilibrium state and ionic current-voltage characteristic of MOS structure, is presented.

The whole book is divided into 7 chapters. After introducing the subject in the first chapter, Chap. 2 deals with the background studies of the MOS structure ideal and non-ideal case. Chapter 3 presents methods typically used to grow oxide such as thermal oxidation, chemical oxidation (anodic oxidation) and Rapid Thermal Oxidation (RTO) as well as oxide-charges and different effects of these charges on the device performance. A complete review of transport Mechanism in thin oxides of MOS devices is discussed in Chap. 4. The studies carried out on mobile ionic charge in thermally oxidized silicon system were primarily aimed at determining its total density. Accordingly, several experimental techniques such as Charge Pumping (CP) technique associated with the BTS method have been developed for measuring the mobile ion concentration in oxides of MOS structures that are described in Chap. 5. Nevertheless, none of these techniques so developed has been used to obtain the density distribution of mobile ions. However, a few theoretical attempts have also been made in this direction, which are reviewed in Chap. 6. In the present work, certain new attempts have been investigated towards the determination of the mobile ionic density-distribution. In the first attempt, methods have been developed to determine the density-distribution of mobile ionic charge in explicit form simply from the knowledge of the measured values of flat band voltage under three different conditions, namely, before contamination/ activation, after contamination/activation, and finally after ion-drift. In other method, a numerical modelling has been used to obtain the mobile ion profile not only at equilibrium state but also its profile evolution when BTS is applied. These methods are presented in Chap. 6. However, the density-distribution of mobile ions in the oxide has been obtained analytically which is based on the argument that these ions must attain an equilibrium density-distribution under the influence of various internal and external forces which are acting upon them. This analytical model of the density-distribution of mobile ions is described in Chap. 7. In the other attempt, a new approach of determining dynamic ionic current-voltage characteristic that is due to ion transport phenomenon in the oxide is presented in Chap. 7. In that approach, the formulation of I-V characteristics of MOS device can be achieved using the theoretical model of mobile ion distribution in oxides.

This work can be used by device physicists, characterisation engineers, or any researches interested in the studies of the MOS device properties.

Boumerdes, August 2010   Hamid Bentarzi

Contents

Contents

List of Symbols, Abbreviations and Physical Constants

A	Area
A_{G}	Area of the channel
C	Capacitance
C_{D}	Depletion capacitance
C_{HF}	High frequency capacitance
C_{it}	Interface trap state capacitance
C_{LF}	Low frequency capacitance
C_{OX}	Oxide capacitance
C_{Si}	Silicon capacitance
C_{MOS}	MOS capacitance
D	Diffusion coefficient
D_{it}	Interface trapped charge density
E	Energy
E_{A}	Activation energy
E_{c}	Conduction band energy
E_{F}	Fermi energy
E_{g}	Energy bandgap
E_{ae}	Activation energy of electrons
E_{d}	The trap energy
E_{v}	Valence band energy
E_{o}	Activation energy at the interface
F	Flux, force
f	Frequency
G	Conductance
G_{MOS}	Leakage conductance through MOS structure
h	Planck constant
J	Current density
J_{DT}	Current density in direct tunneling
J_{FN}	Current density in fowler–nordheim tunneling

k	Boltzmann coefficient
L_D	Debye length
m	Mass
m_o	Free mass of electron
m^*	Effective electron mass
m_{Si}	Effective electron mass in the silicon
m_{ox}	Effective electron mass in the oxide
n	Charge density
N_c	Density of states in the oxide conduction band
n_i	Intrinsic carrier concentration
n_s	Electron density at the silicon/silicon dioxide interface
n_A	Density of ionized acceptors
n_D	Density of ionized donors
N_t	Trap density
N_m	Mobile ions density
N_{ox}	Oxide charge density
N_s	Silicon/Silicon dioxide interface charge density
P	Hole density
q	Unit charge
Q	Charge
Q_{it}	Interface trapped charge
Q_f	Fixed oxide charge
Q_{ot}	Oxide trapped charge
Q_m	Mobile ionic charge
Q_{tot}	Total charge
R	Resistance
t	Time
t_{ox}	Oxide thickness
T	Temperature in kelvin
V	Voltage
V_A	Applied voltage
V_F	Effective gate voltage
V_{FB}	Flat band voltage
ΔV_{FB}	Flat band voltage shift
V_G	Gate voltage
V_{ox}	Voltage across the oxide
V_{Th}	Threshold voltage
W_{ms}	Work function difference between the metal and S_i
x	Distance from metal–oxide interface
X	Centroid of the charge distribution
ε_o	Permittivity of the free space
ε_{si}	Relative permittivity of the silicon
ε_{ox}	Relative permittivity of the oxide
ϕ	Potential
ϕ_B	Bulk potential

ϕ_M	Barrier height at metal/oxide interface
ϕ_S	Surface potential
ϕ_{Si}	Barrier height at silicon/oxide interface
$\psi(x)$	Band bending
ψ_S	Total band bending
ζ	Electric field
ζ_s	Electric field at the silicon/silicon dioxide interface
ζ_{ox}	Electric field in the silicon dioxide
μ	Mobility
μ_e	Mobility of electron
μ_{ion}	Mobility of ion
ρ	Charge density

BTS	Bias thermal stress
CP	Charge pumping
CV	Capacitance voltage
CVD	Chemical vapor deposition
FET	Field effect transistor
FN	Fowler nordheim
HF	High frequency
IC	Integrated circuit
LF	Low Frequency
LPCVD	Low pressure CVD
MOS	Metal oxide semiconductor
MOSFET	Metal oxide semiconductor field effect transistor
NMOS	N-Channel MOS
PECVD	Plasma enhanced CVD
PMOS	P-Channel MOS
RF	Radio frequency
RTD	Resonance tunneling diode
RTA	Rapid thermal annealing
RTO	Rapid thermal oxidation
SCL	Space charge layer
TSIC	Thermally stimulated ionic current

ε_o	8.85×10^{-12} F/m
ε_{ox}	4.1
ε_{si}	11.8
μe	1400 cm^2/(Vs), in Si
k	1.4×10^{-23} J/K
h	$6.62606896 \times 10^{-34}$ Js
q	1.6×10^{-19} Coulomb

Chapter 1
Introduction

In the past decades, the study of the MOS structures has been of great importance to the development of integrated circuit technologies. The motivation behind the use of the silicon dioxide has been the fabrication of stable and high performance MOS devices and integrated circuits. The silicon dioxide that has electrically isolated the transistor gate from the silicon channel is a key material for the digital revolution with today's GHz microprocessors. It is ideally suited to its role, meeting demands for high performance (speed), low static (off-state) power and a wide range of power supply and output voltages. This dielectric currently enables defect charge densities about 10^{10} cm^{-2} eV, and high breakdown fields in excess of 10 MV/cm [1].

Earlier attempts to fabricate MOS devices were unsuccessful because of the lack of controllable and stable surface. Until 1960, stable and reproducible solid–solid interface between silicon and a grown oxide film of known composition and structure could not be produced. Brown [2] in 1953, and Garret and Brattain [3] in 1955 formulated theoretical modeling of surface band bending and its consequences. This theoretical background was applied in the next decade to the silicon-oxide interface. Ligenza [4] made the first good quality oxide films of high dielectric breakdown strength and low loss in high-pressure steam during the years 1960–1961.

A major breakthrough in semiconductor processing occurred in the early 1960, with the development of the silicon planar process that was described first by Hoerni [5]. However, attempts at MOSFET fabrication were blocked by charge migration problems that led to deterioration of their electrical characteristics. In the period 1963–1964, various charges associated with the thermally oxidized silicon structure were observed to cause serious yield and reliability problems. Subsequently, a number of investigations concerning oxide charges were started in various laboratories and many have continued to the present time [6]. During that period, it has been generally established that four general types of oxide charges are associated with the MOS system [7, 8]. These four general types are interface-trapped charge, fixed

H. Bentarzi, *Transport in Metal-Oxide-Semiconductor Structures*,
Engineering Materials, DOI: 10.1007/978-3-642-16304-3_1,
© Springer-Verlag Berlin Heidelberg 2011

oxide charge, oxide-trapped charge and mobile ionic charge. Initially, most of the studies were devoted to the processing parameters so that their adverse effects on device properties could be minimized. Recently, efforts have been focused on a quantitative understanding of the densities, cross sections, and the nature of the oxide charges and traps so that ultimate device performance might be achieved. With further process refinements, the first reliable and reproducible discrete MOSFET's and simple integrated circuits (IC's) were produced on commercial basis in that period.

The serious problem of instability is caused in the devices by mobile ionic oxide charge which is commonly quantified by using a MOS capacitor and by measuring the flat-band voltage shift (ΔV_{FB}) after a Bias Thermal Stress (BTS) test has been applied to it. The mobile oxide charge is not related to the structure of the $Si–SiO_2$ system. It is just due to impurity ions introduced during or after processing. This means that in principle mobile oxide charge can be avoided if care is taken to prevent these impurities from reaching the device (during or after processing).

Back in 1965, Snow et al. [9] determined the shift in the flat-band voltage, which is due to the motion of alkali ions in SiO_2. At that time, one knew from conductivity studies that the alkali atoms are so abundant everywhere that their ions could easily be introduced and incorporated in layers of oxide grown on silicon.

A few decades ago, the amount of contamination was so high that a stable device could not be manufactured and thus the other types of charges could not be investigated. Nowadays, these contamination problems have been greatly reduced by using guttering and very "clean" processing techniques. In laboratories, it is somehow believed that mobile oxide charge no longer plays an important role in the observed instabilities of semiconductor devices and that mobile ions in SiO_2 are only of historical importance. This contention has been now disproved and the importance of the study of mobile ions in the oxides has again been revived. It is, because, not only the initial mobile ion contamination but also the activation of the already existing neutral ions of the device during the subsequent processing stages which can cause serious problems in device performance.

Since very clean processing conditions are not automatically present in all laboratories and are achieved only as a result of extra investment, experience and care during processing, ordinarily the level of ionic contamination may be too high and one has to know where to look for improvements [10].

The use of guttering techniques to render mobile ions to inactive form has several drawbacks that become more apparent with the reduction of the dimensions of modern devices. In such cases, the quality of the SiO_2 layer and of the $Si–SiO_2$ interface becomes of greater importance. Besides, the importance of the study of mobile ionic charge can be further established when intentionally contaminated MOS devices are used as tool in the device characterization. In this way, more insight has been gained in fields such as:

- the structure of the Si–SiO$_2$ interface,
- the interaction between the electrodes material and SiO$_2$ during low temperature anneal processes,
- the wear-out and breakdown phenomena.

More recent studies of the electron trapping on N$^+$-related traps and that of the relation between mobile ions and oxide defects have been carried out. Using samples with very low contamination levels, a kind of interface spectroscopy can be performed with extremely high sensitivity.

Even in samples containing initially very low concentrations of mobile ions, Na$^+$ contamination can be enhanced under the influence of exposure of ions, X-rays, and laser beams to the SiO$_2$ layer. As these types of beams play an increasingly important role in the present and future processes, instability due to the resulting mobile Na$^+$ ions may be anticipated.

"Dry" process steps, those using plasmas, are more prone to pollution than first thought. They can contribute to the ionic contamination of all insulating layers that are etched. Although it does not usually affect the quality of the thermal oxide in contact with the silicon substrate, this contamination is of paramount importance when one considers the total charge distributed in the insulating layer.

Recent efforts in gate dielectric scaling have focused on extending the use of SiO$_2$ through thinning or introducing nitrogen into the dielectric. Moreover, the apparent robust nature of thin SiO$_2$, coupled with the industry's acquired knowledge of oxide process control, has extended its use for past several decades in CMOS technology. Even recently, it has been demonstrated that transistors with gate oxides as thin as 13–15 Å continue to operate satisfactorily [11, 12].

The present work is devoted to the study of mobile ions and their effects in the oxide of MOS structure. After introducing the subject in the present chapter, next chapters deal with the background studies of the MOS structure ideal and non-ideal case. The studies carried out in our laboratory on mobile ionic charge in thermally oxidized silicon system were primarily aimed at determining its total density using an experimental technique such as Charge Pumping (CP) technique associated with the BTS method [13] that is described in Chap. 5. Besides, certain theoretical attempts have also been investigated in our laboratory towards the determination of the mobile ionic density-distribution. In the first attempt, methods have been developed to determine the density-distribution of mobile ionic charge in explicit form simply from the knowledge of the measured values of flat-band voltage under three different conditions, namely, before contamination/activation, after contamination/activation, and finally after ion-drift [14]. In the other method, a numerical modeling has been used to obtain the mobile ion profile at not only equilibrium state but also its profile evolution when BTS is applied. However, in last attempt, the density-distribution of mobile ions in the oxide has been obtained analytically which is based on the argument that these ions must attain an equilibrium density-distribution under the influence of various internal and external forces that are acting upon them [15]. This analytical model has then been used in determining dynamic ionic current–voltage (I–V) characteristic that is due to ion transport phenomenon in the oxide. Finally, the I–V characteristics of MOS device have been computed and compared with experimental I–V curves [16].

References

1. Wallance, R.M., Wilk, G.D.: Exploring the limits of gate dielectric scaling. Semicon. Intern. J. **24**, 153–158 (2001)
2. Brown, W.L.: n-Type conductivity on p-type germanium. Phys. Rev. **91**, 518–527 (1953)
3. Garrett, C.G.B., Brattain, W.H.: Physical Theory of Semiconductor Surfaces. Phys. Rev. **99**, 376–387 (1955)
4. Ligenza, J.R.: Effect of crystal orientation on oxidation rates of silicon in high pressure steam. J. Phys. Chem. **65**, 2011–2014 (1964)
5. Hoerni, J.A. : Planar silicon diodes and transistors. IRE Trans. Elect. Dev. **ED-8**, 178 (1961). Also presented at Professional Group on Electron Devices Meeting, Washington, D.C., October. (1960)
6. Agajanian, A.H.: Semiconductor Devices. A Bibliography of Fabrication Technology, Properties, and Applications. Plenum, New York (1976)
7. Cheng, Y.C.: Electronic states at the silicon–silicon dioxide interface. Prog. Surf. Sci. **8**, 181–218 (1977)
8. Deal, B.E.: The current understanding of charges in the thermally oxidized silicon structure. J. Electrochem. Soc. **121**, 198–205 (1974)
9. Snow, E.H., Grove, A.S., Deal, B.E., et al.: Ion transport phenomena in insulating films. J. Appl. Phys. **36**, 1664–1673 (1965)
10. Hillen, M.W., Verwey, J.F.: Mobile ions in SiO_2 layers on Si. In: Barbottain, G., Vapaille, A. (eds.) Instabilities in Silicon Devices, pp. 404–439. Amsterdam (1986)
11. Timp, G., et al.: The ballistic nano-transistor. IEEE IEDM Tech. Dig. pp. 55–58 (1999)
12. Weir, B.E., et al.: Gate oxide in 50 nm devices: thickness uniformity improves projected reliability. IEEE IEDM Tech. Dig. pp. 437–440 (1999)
13. Bentarzi, H., Zitouni, A., Kribes, Y.: Oxide charges densities determination using charge-pumping technique with BTS in MOS structures. WSEAS Trans. Electron. 101–110 (2008)
14. Bentarzi, H., Bouderbala, R., Mitra, V.: Determination of the distribution of mobile charges in the oxide of the MOS structure. ESD'94. II, Brno. 106–111 (1994)
15. Mitra, V., Bentarzi, H., Bouderbala, R., Benfdila, A.: A theoretical model for the density-distribution of mobile ions in the oxide of the metal-oxide-semiconductor structures. J. Appl. Phys. **73**, 4287–4291 (1993)
16. Bentarzi, H., Bouderbala, R., Zerguerras, A.: Ionic current in MOS structures. Ann. Telecommun. **59**(3–4), 471–478 (2004)

Chapter 2
The MOS Structure

2.1 Introduction

The metal-oxide-semiconductor diode or MOS capacitor is an important structure, which is incorporated in the surface of most semiconductor devices. It forms an essential part of a MOSFET which in turn is an important device used in large-scale integration. Therefore, all the studies related to any kind of MOS device needs at first the basic understanding of the MOS structure. In order to achieve this objective, the present chapter is devoted to the study of MOS structure. A simple physical approach applied to MOS structure and a behavior of ideal MOS capacitor [1–5] that are necessary for understanding the analyses that will follow subsequently, are described. At first, all the basic concepts and quantities are introduced. Then, the charge distribution that sets in a MOS structure when the latter is biased in either one of the three biasing modes (accumulation, depletion, and inversion) is analyzed. This charge distribution is used to obtain the value of the capacitance and its dependence on the magnitude and the frequency of the applied small signal using a phenomenological approach. A real MOS structure always contains so-called "oxide charges" located in the bulk of the oxide or at the oxide-silicon interface. The impact of these charges on the behavior of real MOS structure and in particular on the flat-band voltage is also examined.

2.2 A Simple Physical Approach Applied to MOS Structure

The MOS capacitor consists of an oxide film sandwiched between a P- or N-type silicon substrate and a metal plate called gate as shown in Fig. 2.1. The study of the behavior of this capacitor under a varying bias applied between substrate and gate is a powerful way to investigate the quality of the oxide layer and the quality of the oxide-silicon interface.

H. Bentarzi, *Transport in Metal-Oxide-Semiconductor Structures*,
Engineering Materials, DOI: 10.1007/978-3-642-16304-3_2,
© Springer-Verlag Berlin Heidelberg 2011

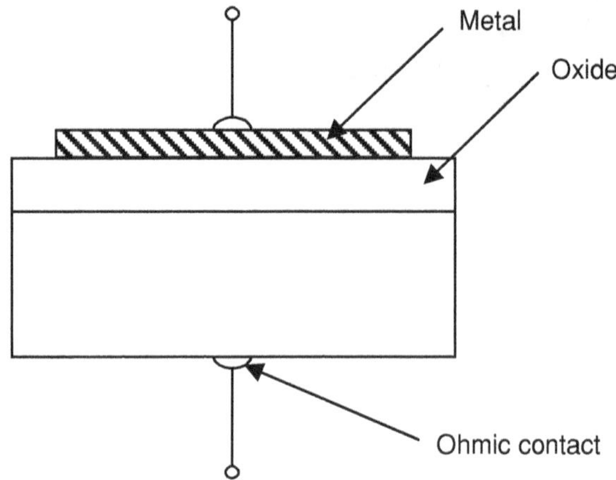

Fig. 2.1 Cross-sectional view of a MOS structure

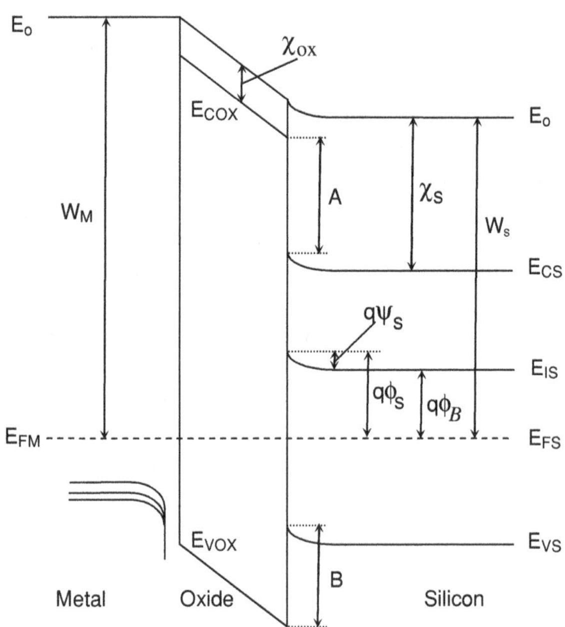

Fig. 2.2 Energy band diagram of unbiased real MOS structure ($W_M \neq W_S$)

2.2.1 Basic Concepts and Quantities

Figure 2.2 shows the energy band diagram of an unbiased MOS structure when the work function of the metal W_M and work function of silicon W_S are different. The diagram shows the position of the different energy levels such as Fermi level in the gate (E_{FM}) and in the silicon (E_{FS}). In this figure, χ_S represents the electron affinity for the silicon and χ_{ox} for the oxide. Figure 2.2 also shows that certain energy barriers exist between the metal and the oxide as well as between the silicon and

Table 2.1 Some typical values in the energy bands of a MOS structure

Metal	Oxide	Silicon
$W_M = 4.8$ eV (Au)	$\chi_{ox} = 0.9$ eV	$\chi_S = 4.1$ eV
$W_M = 4.1$ eV (Al)	E_{COX}–$E_{VOX} = 8.1$ eV	E_{CS}–$E_{VS} = 8.1$ eV
	$A = 3.2$ eV	4.1 eV $< W_S^* < 5.2$ eV
	$B = 3.8$ eV	

*W_S varies with doping concentration and temperature

the oxide. For example, an energy $(W_M$–$q\chi_{ox})$ would be needed to move an electron from the Fermi level of the metal E_{FM} to the lowest unoccupied states in the oxide, and $A + (E_{CS}$–$E_{VS})$ would be needed to move an electron from the silicon valence band to the lowest unoccupied states in the oxide, where W_M is the work function of metal, E_{CS} and E_{VS} the bottom of conduction band and top of valence band of silicon respectively. "A" difference between the bottom level of the conduction bands of oxide and silicon at the Si–SiO$_2$ interface and q the electron charge. The importance of these energy barriers is that they prevent the free flow of carriers from the metal to the silicon or vice versa. Some typical values for such a structure are shown in Table 2.1 [6–8].

2.2.2 Definition of Potentials

Figure 2.2 shows the various potentials. The potential may be defined by the following equation,

$$q\phi = E_F - E_i(x). \tag{2.1}$$

where E_F is the extrinsic Fermi level and E_i is the intrinsic energy level in the silicon. The potential $\phi(x)$ is called the *bulk potential* ϕ_B in the bulk $(x \to \infty)$ and the *surface potential* ϕ_S at the surface $(x = 0)$.

Location of any other energy level e.g. an interface trap level within the silicon band gap may be specified by stating its distance in electron volt from the intrinsic level. The *band bending* $\psi(x)$ is defined as:

$$\psi(x) = \phi(x) - \phi_B. \tag{2.2}$$

where $\psi(x)$ represents the potential at any point x in the depletion layer with respect to its value in the bulk. In particular, the barrier height $\psi_S = \phi_S - \phi_B$ is the *total band bending*.

2.3 Ideal MOS Capacitor

Before characterizing electrically the real MOS device by taking into consideration the defects contained in the SiO$_2$, at first, the ideal MOS structure will be studied. The MOS structure is called ideal if the following two conditions are met:

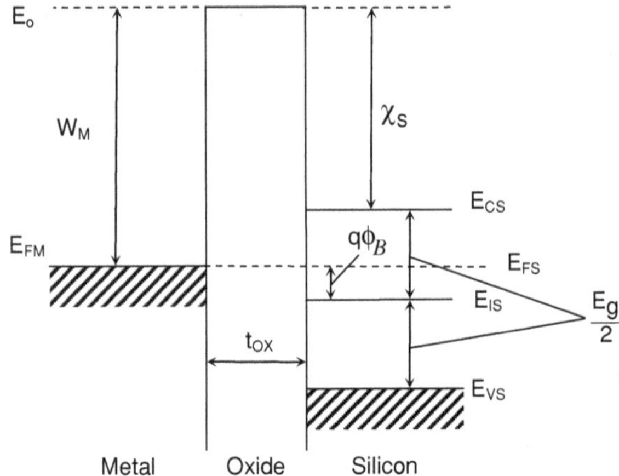

Fig. 2.3 Energy band diagram of unbiased ideal MOS structure

1. The work function of metal W_M and work function of silicon W_S are equal, $W_M = W_S$, which implies that in the three materials, all energy levels are flat, when no voltage applied to the structure. This case is illustrated in Fig. 2.3.
2. There exists no charge in the oxide and at the Si–SiO$_2$ interface, which implies that the electric field is zero everywhere in the absence of any applied voltage.

MOS capacitance will vary with the applied gate to substrate voltage. The capacitance versus voltage characteristics of MOS capacitors that result from the modulation of the width of the surface space charge layer (SCL) by the gate field have been found to be extremely useful in the evaluation of the electrical properties of oxide-silicon interfaces. There are three regions of interest, namely, accumulation, depletion and inversion in the C–V characteristics of the MOS capacitor as shown in Fig. 2.4. A MOS capacitor fabricated on a P-type substrate is the case treated here.

2.3.1 Accumulation

When an external voltage V_G is applied to the silicon surface in MOS capacitor, the carrier densities change accordingly in its surface region. With large negative bias applied to the gate, holes are attracted by the negative charges to form an accumulation layer (Fig. 2.5). The high concentration of these holes will form the second electrode of a parallel plate capacitor with first electrode at the gate. Since the accumulation layer is an indirect ohmic contact with the P-type substrate, the capacitance of the structure under accumulation conditions must be approximately equal to the capacitance of the oxide [1],

$$C_{ox} = \frac{\varepsilon_o \varepsilon_{ox}}{t_{ox}} \tag{2.3}$$

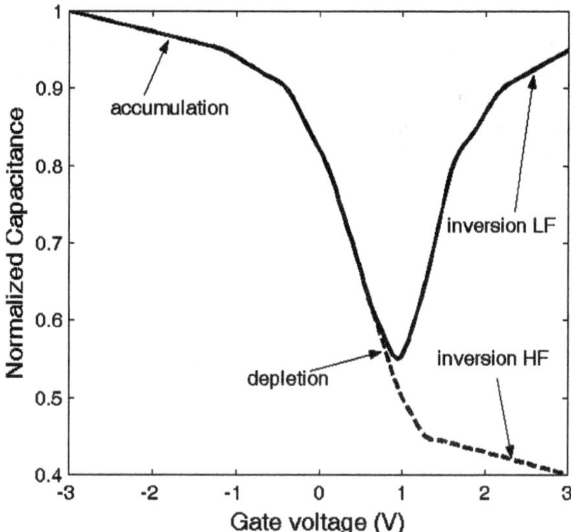

Fig. 2.4 Typical ideal C–V curves showing the three modes: accumulation, depletion, inversion for both high and low frequency in a P-MOS structure

where ε_o is the permitivity of the free space, ε_{ox} the relative permitivity of oxide, and t_{ox} the oxide thickness. This capacitance is always expressed per unit gate area [F cm^{-2}]. It does not vary with bias V_G as long as the structure is maintained in accumulation mode (Fig. 2.4). It is also independent of the frequency as long as the motion of the majority carriers, which contribute to substrate charge ΔQ_S, can keep pace with the incremental speed of gate charge ΔQ_M. This is true if the frequency of the applied small signal is smaller than the reciprocal of the dielectric time constant of silicon, i.e. 10^{11} Hz. Under this condition, the Fermi level near the silicon surface will move to a position closer to the valance band edge as shown in Fig. 2.5c.

2.3.2 Depletion

When negative charges are removed from the gate, holes leave the accumulation layer until the silicon will be neutral everywhere. This applied gate bias is called the flat band voltage. As the bias on the gate is made more positive with respect to flat band, holes are repelled and a region is formed at the surface which is depleted of carriers (Fig. 2.6b). Under depletion conditions, the Fermi level near the silicon surface will move to a position closer to the center of the forbidden region as illustrated in Fig. 2.6c. Increasing the positive voltage V_G will tend to increase the width of the surface depletion region X_D, the capacitance from the gate to the substrate associated with MOS structure will decrease, because the capacitance associated with the surface depletion region will add in series to the capacitance across the oxide. Thus the total capacitance per unit area from the gate to substrate under depletion conditions is given by

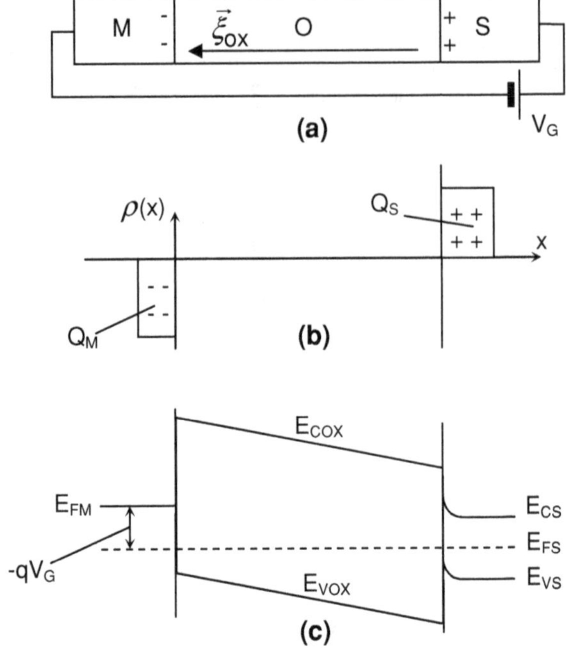

Fig. 2.5 Schematic representation of P-MOS structure under bias resulting in accumulation mode, **a** biasing condition, **b** charge distribution, **c** energy band diagram

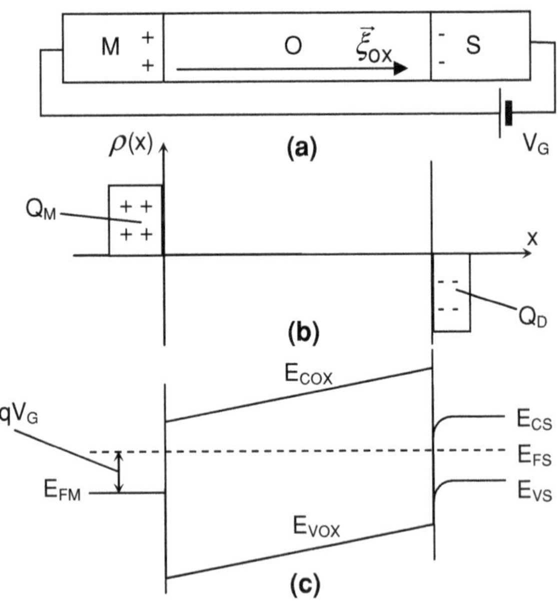

Fig. 2.6 Schematic representation of P-MOS structure under bias resulting in depletion mode, **a** biasing condition, **b** charge distribution, **c** energy band diagram

$$C(V_G) = \left(\frac{1}{C_{ox}} - \frac{1}{C_S(V_G)} \right)^{-1}, \tag{2.4}$$

where C_S is the silicon capacitance per unit area, is given by

$$C_S(V_G) = \frac{\varepsilon_o \varepsilon_S}{X_D}, \tag{2.5}$$

and,

$$X_D = \sqrt{\frac{2\varepsilon_o \varepsilon_S \psi_S}{qN_A}}. \tag{2.6}$$

Where the relation between the applied gate voltage V_G and the total band banding ψ_S can be written as

$$V_G = \psi_S + \frac{\sqrt{2\varepsilon_o \varepsilon_S qN_A \psi_S}}{C_{ox}} \tag{2.7}$$

Since only majority carriers contribute to the substrate charge ΔQ_D, the capacitance is independent of frequency.

2.3.3 Inversion

With increasingly applying positive voltage, the surface depletion region will continue to widen until the onset of surface inversion is observed (n-type), an inversion layer is formed, the Fermi level near the silicon surface will now lie close to the bottom of conduction band (Fig. 2.7). This inversion layer is very thin (1–10 nm) and separated from the bulk of silicon by the depletion layer. The build-up of inversion layer is a threshold phenomenon. The threshold condition marks the equality of the concentration of minority carriers to the doping concentration. At the onset of inversion, the depletion layer width reaches a limit, X_{DLim} as shown in Fig. 2.7b. Since the charge density in the inversion layer may or may not be able to follow the ac variation of the applied gate voltage, it follows that the capacitance under inversion conditions will be a function of frequency.

Low frequency Capacitance

This case, illustrated in Fig. 2.4, corresponds to the thermal equilibrium in which the increase in the gate charge δQ_M is balanced by the substrate charge δQ_{inv}. It arises when the frequency of the small signal is sufficiently low (typically less than 10 Hz). The low frequency capacitance of the structure, C_{LF}, is equivalent to that of the oxide layer, just as in accumulation mode,

$$C_{LF} = C_{ox}. \tag{2.8}$$

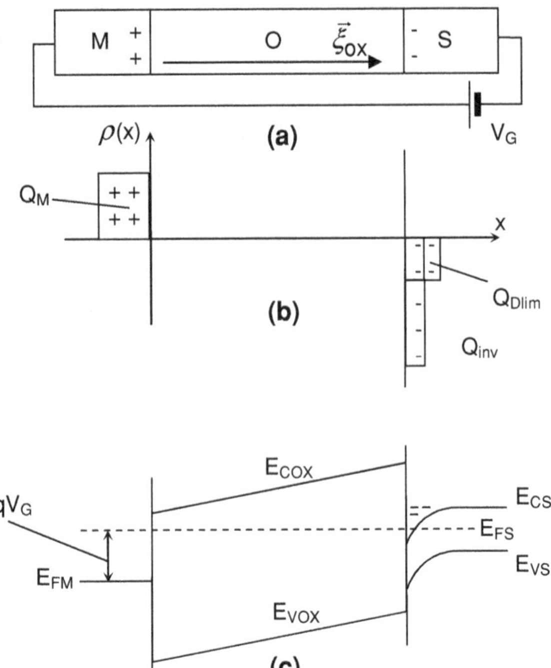

Fig. 2.7 Schematic representation of P-MOS structure under bias resulting in inversion mode, **a** biasing condition, **b** charge distribution, **c** energy band diagram

High Frequency capacitance

The case illustrated in Fig. 2.4, corresponds to the higher frequencies of the applied small signal (typically above 10^5 Hz). The increase of charge in the metal side δQ_{M} is now balanced by the substrate charge δQ_{D}, since the minority carriers can no longer adjust their concentrations. The charge modulation δQ_{D} occurs at distance X_{DLim} of the Si–SiO$_2$ interface. It follows that the high frequency capacitance of the MOS structure, C_{HF}, is given,

$$\frac{1}{C_{\mathrm{HF}}} = \frac{1}{C_{\mathrm{ox}}} + \frac{1}{C_{\mathrm{D\,lim}}}. \tag{2.9}$$

where

$$C_{\mathrm{D\,lim}} = \frac{\varepsilon_o \varepsilon_S}{X_{\mathrm{D\,lim}}}, \tag{2.10}$$

and,

$$X_{\mathrm{D\,lim}} = \sqrt{\frac{4\varepsilon_o \varepsilon_S kTLn\left\langle \frac{N_{\mathrm{A}}}{n_{\mathrm{i}}} \right\rangle}{q^2 N_{\mathrm{A}}}}. \tag{2.11}$$

As shown in Fig. 2.4, the capacitance is practically independent of positive or negative bias for both high frequency inversion and low frequency inversion.

2.4 The Actual (Non-ideal) MOS Structure

An ideal MOS device does not agree with experimental results, and this difference is due to the presence of the oxide charges and the work function difference that exists in practice but was not taken into account in the theoretical treatment of an ideal MOS capacitor. Early studies of the MOS devices showed that the threshold voltage V_{Th} and the flat band voltage V_{FB} could strongly be affected by these charges. The understanding of the origin and nature of these charges is very important if they are to be controlled or minimized during device processing [2, 9]. The net result of the presence of any charge in the oxide is to induce a charge of opposite polarity in the underlying silicon. The amount of charge induced will be inversely proportional to the distance of the charge from the silicon surface. Thus, an ion residing in the oxide very near the Si–SiO$_2$ interface will reflect all of its charge in the silicon, while an ion near the oxide outer surface will cause little or no effect in the silicon. The charge is measured in terms of the net charge per unit area at the silicon surface. Most oxide charge evaluations can be made using the capacitance voltage (C–V) method. This method is simple and rapid [10, 11] and in most cases provide a quantitative or at least a semiquantitative measure of the surface charge.

2.4.1 The Metal-Silicon Work Function Difference

In the real MOS structure, the work function of the metal and the work function of the silicon are different [6, 7]. For this reason, there exists an electric field in the oxide and in the top layer of the silicon even in the absence of an applied voltage (see band diagrams of Fig. 2.2). To obtain the flat band conditions, $\psi_S = 0$, a bias on the gate must be applied relative to the substrate, which can be written as

$$\Delta V_{FB1} = \frac{W_{MS}}{q}. \qquad (2.12)$$

As an example, for Al–SiO$_2$–Si structure, a typical value of ΔV_{FB1} is 0.3 V for an n-type Si substrate and 0.8 V for a p-type [6]. The effect of a work function difference may cause a shift of the actual C–V curve with respect to the ideal one. The flat-band-voltage shift ΔV_{FB1} occurs along the voltage axis as illustrated in Fig. 2.8.

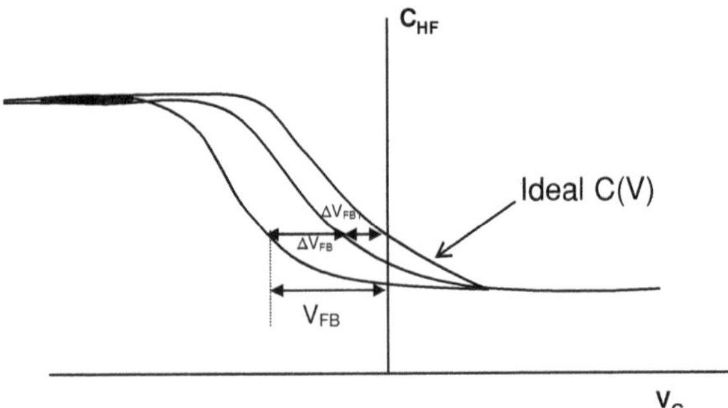

Fig. 2.8 Representation of the C–V curves of a P-type MOS structure showing the flat band voltage shifts introduced by work function difference and oxide charges

2.4.2 Effect of the Charge Distributed in the Oxide

Whether mobile ions or other types of oxide charges are distributed unevenly in the bulk, their density $\rho(x)$ varies with distance (and with time in case of time-dependent stress). To study the influence of oxide charges distribution on the properties of the MOS structure, at first, the effect of only those charges, which are located within a layer between x and $x + dx$, is calculated. The origin of the x-axis is taken at the metal-oxide interface as shown in Fig. 2.9. In a second step, the effect of the various layers from zero to t_{ox} is added. Using Gauss's law, the electric field in the oxide ζ_{ox} exhibits a discontinuity $\delta\zeta_{\mathrm{ox}}$ when crossing this charge layer. This discontinuity is given by

$$\delta\zeta_{\mathrm{ox}} = \frac{\rho(x)dx}{\varepsilon_0\varepsilon_{ox}}. \tag{2.13}$$

For ensuring flat band condition in the silicon, ζ_{ox} must be zero on the right hand side of the discontinuity. Thus, the profile of the electric field should be as shown in Fig. 2.9b and the corresponding gate voltage that ensures the flat band condition is given by:

$$\delta V_{\mathrm{FB}} = \frac{-\rho(x)xdx}{\varepsilon_0\varepsilon_{ox}} \tag{2.14}$$

Using a classical result of electrostatics, namely the superposition theorem, the effects of all layers comprised between zero and t_{ox} are added and the gate voltage shift ΔV_{FB}, which is necessary to ensure a flat-band condition at the Si–SiO$_2$ interface, is found to be

$$\Delta V_{\mathrm{FB}} = -\int_a^{t_{ox}} \frac{\rho(x)xdx}{\varepsilon_0\varepsilon_{ox}}. \tag{2.15}$$

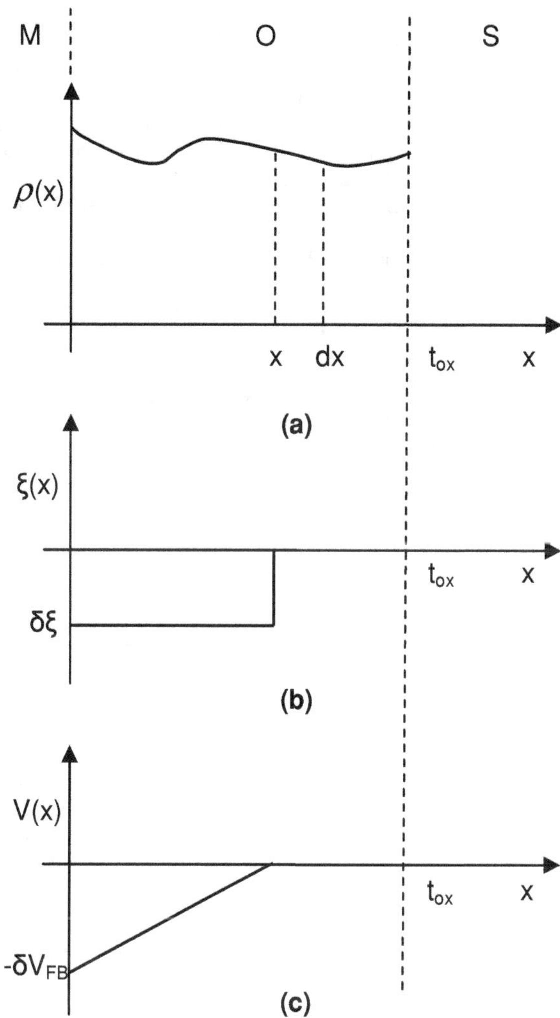

Fig. 2.9 Distribution of **a** oxide charges, **b** electric field and **c** voltage within the oxide of MOS structure

The effect of each charge layer depends on its distance from the oxide-silicon interface as given in Eq. (2.15). A layer has no effect if it is located at the metal-oxide interface and has a maximum effect if it is located at the oxide-silicon interface.

References

1. Nicollian, E.H., Brews, J.R.: MOS Physics and Technology. Wiley, New York (1982)
2. Goetzberger, A., Sze, S.M.: Metal-insulator-semiconductor (MIS) physics. In: Wolfe, R. (ed.) Applied Solid State Science. Academic Press, New York (1969)

3. Grove, A.S.: Physics and Technology of Semiconductor Devices. Wiley, New York (1967)
4. Richman, P.: MOS Field-Effect Transistors and Integrated Circuits. Wiley, New York (1973)
5. Prezewlocki, H.M.: Work function difference in MOS structures: current understanding and new measurement methods (1982)
6. Maykusiak, B., Jakubowski, A.: A new method for the simultaneous determination of the surface-carrier mobility and the metal-semiconductor work-function difference in MOS transistors. IEEE Trans. Elect. Dev. **ED-35**, 439–443 (1988)
7. McNutt, M.J., Sah, C.T.: Determination of the MOS oxide capacitance. J. Appl. Phys. **46**, 3909–3913 (1975)
8. Deal, B.E.: Standardized terminology for oxide charge associated with thermally oxidized silicon. IEEE Trans. Elect. Dev. **ED-27**, 606–608 (1980)
9. Deal, B.E.: The current understanding of charges in the thermally oxidized silicon structure. J. Electrochem. Soc. **121**, 198–205 (1974)
10. Terman, M.: An investigation of surface states at silicon–silicon oxide interface employing metal oxide silicon diodes. Solid St. Elect. **5**, 285–299 (1962)
11. Snow, E.H., Grove, A.S., Deal, B.E., et al.: Ion transport phenomena in insulating films. J. Appl. Phys. **36**, 1664–1673 (1965)

Chapter 3
The MOS Oxide and Its Defects

3.1 Introduction

The insulator layer so-called oxide film is an important part that is incorporated in the MOS structure that is in turn an important device used in large-scale integration. The basic MOS structure consists of an oxide film (silicon dioxide or silicon dioxide and silicon nitride layers) sandwiched between a metal (often aluminum or heavily doped Poly-silicon layer) and P- or N-type silicon substrate. This oxide film can be grown using different techniques such as thermal oxidation, chemical oxidation (anodic oxidation) and Rapid Thermal Oxidation (RTO).

A real MOS oxide always contains defects so-called "charges" located in the bulk of the oxide or at the oxide–silicon interface as shown in Fig. 3.1. These charges are traditionally classified into four general types as proposed by a committee of scientists in 1980 [1]:

- Fixed oxide charge,
- Mobile ionic charge,
- Interface-trapped charge,
- Oxide trapped charge.

Their nomenclature, location, and electric properties are discussed along with the techniques that are used to detect them. Besides, the impact of these charges on the behavior of real MOS structure particularly on the flat-band voltage is also studied.

3.2 Oxide Growth Techniques

Silicon dioxide acts as an insulator when grown between two metal layers in a device. A thick oxide (3,000–10,000 Å), referred to as a field oxide, prevents electrical charges from passing between different metal layers and minimizes the

H. Bentarzi, *Transport in Metal-Oxide-Semiconductor Structures*,
Engineering Materials, DOI: 10.1007/978-3-642-16304-3_3,
© Springer-Verlag Berlin Heidelberg 2011

Fig. 3.1 Names and
locations of oxide charges
in the MOS structure

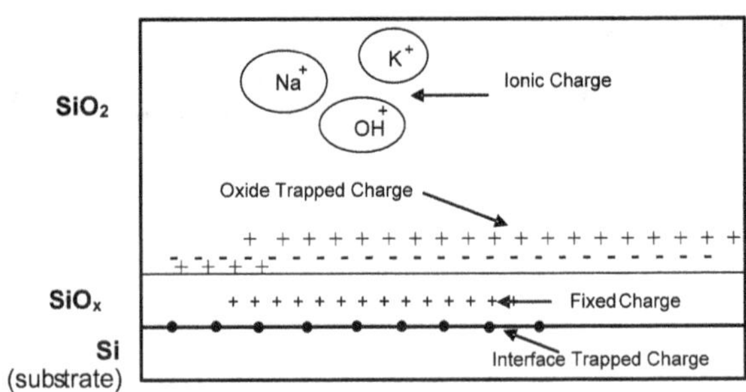

chance for an electrical short circuit to occur in the MOS transistor. A thin layer of
oxide (150–500 Å) is grown in the gate regions of a MOS transistor. A thin oxide
layer still acts as a dielectric, but allows a small electrical charge to pass between
the gate metal and the silicon. This charge called an inductive charge opens up the
gate to allow an electrical current to flow between the source and drain. Tunneling
oxides are thin oxide layers (<100 Å) grown between super-conducting materials
in memory devices. Due to the thinness of the oxide, electrons can pass through it
with zero resistance.

The goal of oxidation is to grow a uniformly high quality oxide layer on a
silicon substrate. When the oxidant is oxygen, the reaction with silicon produces
silicon dioxide. As the silicon dioxide layer grows, the chemical reaction between
the oxidants and the silicon consumes the silicon atoms. As a general principle, the
amount of silicon consumed in the oxidation reaction is 44% of the final oxide
thickness [2]. For example, growing 10,000 Å of oxide consumes 4,400 Å of
silicon. Oxidation of silicon occurs at room temperature, and despite the slow rate
of oxide growth, a thin layer (<20 Å) of silicon dioxide forms. However, to be cost
effective, manufacturers have developed oxidation techniques that speed up the
silicon dioxide growth rate.

Wafer oxidation involves four steps:

(1) the wafers are placed in furnace,
(2) the wafers are oxidized,
(3) the wafers are removed from furnace and cleaned,
(4) the oxide on the wafer is inspected.

The methods typically used to grow oxide are thermal oxidation, chemical
oxidation (anodic oxidation) and Rapid Thermal Oxidation (RTO). Thermal oxidation is the most common oxidation process used today to grow silicon dioxide
on silicon and is the focus of the next section.

3.3 Thermal Oxidation

Thermal Oxidation refers to an oxide growth process occurring at high temperatures. Growing oxides at high temperatures increases the oxide growth rate,

making oxidation cost effective. During thermal oxidation, silicon is exposed to the oxidants at temperatures between 900 and 1,200°C. Consequently, the rate of oxide growth increases dramatically [3]. With thermal oxidation, oxides can be grown under both wet and dry conditions.

Thermal oxidation can be carried out in either a horizontal or vertical tube furnace. Both furnaces have the same heating systems and operate in the same manner, but differ in their structural orientation. Vertical furnaces are more prevalent in industry today because they require less floor space and eliminate many of the problems associated with horizontal furnaces, namely uneven temperature and gas flow. Regardless of the furnace, the oxidation process is the same. Cleaned wafers are placed in the wafer load station where dry nitrogen (N_2) is introduced into the chamber. The nitrogen prevents oxidation from occurring while the furnace reaches the required temperature. Once the specified temperature in the chamber is reached, the nitrogen gas flow is shut off and oxygen (O_2) is added to the chamber. The source of the oxygen can be gas or water vapor state depending upon the process be used. After the oxidation is completed and the oxide layer is the correct thickness, nitrogen is reintroduced into the chamber to prevent further oxidation from occurring. The wafers are then removed from the chamber and following their inspection, they are ready for further processing.

3.3.1 Dry Oxidation

During dry oxidation, dry oxygen is introduced into the process tube where it reacts with silicon. Since dry oxidation is a slow process, approximately 140–250 Å/h, it is only used in industry to grow thin oxides (<1,000 Å). The reaction during dry oxidation is:

$$\text{Si (solid)} + O_2 \text{ (gas)} \rightarrow SiO_2 \text{ (solid)}$$

In dry oxidation, the amount of water in the processing tube is kept at a minimum. If the water level exceeds 25–50 ppm, the oxidation rate increases and a thick layer of poor quality oxide is produced [3]. Growing thin oxides is important in the manufacture of MOS transistors, MOS gates, and dielectric components of devices. Many of these features are smaller than 1 micron, requiring oxides less than 150 Å thickness. High quality thin oxides are difficult to grow because under normal manufacturing conditions the oxidation growth rate is too fast to control it. Therefore, in order to grow a high quality thin oxide, the oxidation process must be slowed down the oxide growth rate. The dry oxidation process allows control over the growth of thin oxides. Introducing hydrochloric acid (HCl), trichloroethylene (TCE) or tri-chloromethane (TCA) with oxygen into the oxidation tube slows down the oxide growth rate. Other adjustments such as reducing the pressure level and lowering the temperature while increasing the pressure also slow down the oxide growth rate and improve the quality of the oxide. For example, 300 Å of oxide can be grown under high pressure (10 atm) at low temperatures (750°C) in 30 min [4].

3.3.2 Wet Oxidation

The silicon dioxide growth rate is faster during wet oxidation, 1,000–1,200 Å/h. Therefore, wet oxidation is the preferred method to grow thick oxides. During wet oxidation, the silicon atoms react with water vapor to produce silicon dioxide as seen in the following reaction:

$$Si\,(solid) + H_2O\,(vapor) \rightarrow SiO_2\,(solid) + 2H_2$$

During wet oxidation, water vapor is introduced into the heated oxidation tube. Because water molecules are smaller than oxygen molecules in size, they diffuse faster in silicon dioxide [5] and the oxide growth rate increases. As the silicon dioxide forms, it traps the hydrogen atoms within it. These hydrogen atoms are released in subsequent processing steps and do not affect the quality of the oxide. There are several different wet oxidation techniques such as Bubblers, Flash system and Dryox system.

Bubblers

A glass flash referred to as a bubbler contains de-ionized water and is attached to the oxidation tube. The water is heated (90–99°C), and water vapor forms above the de-ionized water level. A carrier gas such as nitrogen is bubbled through the de-ionized water. When this carrier gas passes through the vapor, it will be saturated with water. The vapor travels into the oxidation tube, where with additional heating it turns into steam and oxidation occurs. A consistent oxide growth rate is hard to maintain with the bubbler method because of the difficulties involved in controlling both the amount of water vapor entering the oxidation tube and the temperature of the water. The risk of contamination is also high.

Flash System

A flash system is similar in design to a bubbler. A small amount of de-ionized water is dropped on a heated quartz surface where it instantly turns into steam. A carrier gas moves the steam into the heated oxidation chamber. As with the bubbler, it is very difficult to achieve a constant rate of oxide growth. Unlike a bubbler, however, the flask is never opened in a flash system so the risk of contamination is low.

Dryox System

In a Dryox system, oxygen and hydrogen directly enter a heated oxidation tube. In the heated oxidation tube, the two gases mix and form water as steam. The Dryox system is preferred oxidation method for advanced devices because the oxide growth rate can be precisely controlled. Mass Flow Controllers regulate the gas flow into the tube, insuring uniform oxide growth. Contamination of the oxide is limited since this gas is clean. One major disadvantage of Dryox system is the explosive nature of hydrogen at high temperatures. Safety precautions must be taken to minimize the risk of a hydrogen explosion.

3.4 Anodic Oxidation

Anodic oxidation is also referred to as chemical oxidation. During this process, the wafer is attached to a positive electrode and immersed in a bath of Potassium Nitrate (KNO_3). The submersion tank contains a negative electrode. When a current is applied to the system, oxygen is created and a reaction between the silicon and oxygen occurs. This process is used to grow oxide on wafers that will undergo destructive oxide inspection techniques. Unlike thermal oxidation, the silicon moves to the top of the oxide layer during oxidation and the dopant is drawn to the wafer surface. The location of the dopant near the wafer surface allows the resistivity of the wafer and the dopant concentrations to be measured [6].

3.5 Rapid Thermal Oxidation

Rapid thermal oxidation (RTO) is used to grow thin oxides for MOS gates. The RTO systems can heat and cool the wafer quickly and thus effectively control the thickness of the oxide. RTO systems use radiation heat sources such as microwave units, plasma arc, tungsten halogen lamps and graphite heaters rather than traditional heating methods. The heating systems in RTO heat and cool faster than conventional heating systems, allowing more control over the growth rate of thin oxides. However, the heating systems do not always produce a constant temperature across a batch of wafers, and can result in non-uniform oxides. Despite this problem, RTO is still widely used to grow thin oxides on single wafer [7].

3.6 MOS Oxide Defects

The $Si–SiO_2$ system is formed when a layer of silicon dioxide (SiO_2) is either grown thermally or deposited by chemical vapor deposition (CVD) processes on a silicon substrate. In either case, a thin SiO_x layer is formed at the interface between the two materials. In the Si–SiO$_2$ system, defects are induced either in manufacturing processes or by high-field effects such as hot-carrier injection.

Early studies of the MOS devices have shown that the threshold voltage V_{Th} and the flat band voltage V_{FB} could strongly be affected by these charges. The understanding of the origin and nature of these charges is very important if they are to be controlled or minimized during device processing [8]. The net result of the presence of any charge in the oxide is to induce a charge of opposite polarity in the underlying silicon. The amount of charge induced will be inversely proportional to the distance of the charge from the silicon surface. Thus, an ion residing in the oxide very near the Si–SiO$_2$ interface will reflect all of its charge in the silicon, while an ion near the oxide outer surface will cause little or no effect in the silicon. The charge is measured in terms of the net charge per unit area at the silicon surface. Most of the oxide charge evaluations can be made using the capacitance voltage (C–V) method. This method is simple and rapid [9, 10] and in most cases provides a quantitative or at least a semi-quantitative measure of the surface charge. The oxide charges can be classified into four types as proposed by a committee of scientists [1] in 1980. These types of defects in the Si–SiO$_2$ system are the fixed oxide-charge Q_f, the mobile ionic charge Q_m, the interface-trapped charge Q_{it}, and the oxide-trapped charge Q_{ot}. Figure 3.1 illustrates the defects and their location in the Si–SiO$_2$ system. All these types of oxide charges have been thoroughly studied along with the methods that allow measuring their density as well as their distribution. They are summarized in Table 3.1.

3.6.1 The Interface Trapped Charge

It is found that certain energy states exist on the oxide–silicon interface (called interface trapped charge) which lie within the band gap and are originated due to structure [11, 12], oxidation-induced defects [13, 14] or other defects caused by radiation [15, 16]. Interface trapped charges have been studied over a time longer than the other charges associated with passivated semiconductors. They were even studied in the early days of germanium; yet they are still the least understood. The interface trapped charges derive their name from the fact that the traps exist on the oxide–silicon interface. These charges in the past have been called surface states, fast states, interface states. When charge carriers electron/holes occupy these states, they attribute positive or negative charge to the interface depending upon the type of surface state whether donor or acceptor. The charging or discharging of these states depends upon the applied gate voltage. Nicollian and Goetzberger [17], and Brown and Gray [18] have reported data that indicate that the distribution of interface trapped charges across the silicon band-gap typically exhibits peaks near both band edges with a pronounced and uniform minimum occurring near the center of the forbidden region. Brown and Gray [17] have found that, the observed density of interface trapped charges as a function of silicon substrate orientation decreases in the order (111) > (110) > (100).

Table 3.1 Characteristic of different types of oxide charges

Properties	Types			
	Interface trapped charge	Oxide trapped charge	Fixed trapped charge	Mobile ionic charge
Location	At Si–SiO$_2$ interface	Within the oxide bulk	Near Si–SiO$_2$ interface	Within the oxide bulk
Charges	Positive/ negative	Positive/negative	Positive	Positive
Causes	Structural, Metal impurities	Ionizing radiations, Avalanche injection	Structural	Ionic impurities (Na$^+$, K$^+$...)
Dependence on V_G	It depends on V_G	It does not depend on V_G	It does not depend on V_G	It does not depend on V_G
Charging state	Charged and discharged by V_G	Charged and discharged under specific conditions	Fixed	Fixed below 390°C
Density determination method	LF C–V, Conductance, CP	HF C–V, Photo I–V	HF C–V	BTS, TVS, TSIC
Degradation	Hot-carriers	Radiation and hot-carriers	–	Thermal and electric stress

BTS bias-thermal stress, *TVS* triangular voltage sweep, *CP* charge-pumping, *TSIC* thermally stimulated ionic current

At high frequency, the presence of interface-trapped charges does not modify the value of the overall high frequency capacitance but only the relationship between C and V_G. This is because the interface traps cannot adjust their charge state fast enough to keep pace with the applied small alternative signal. The capacitance of the structure is equivalent to an ideal capacitance. This type of charge is responsible for a shift of the C$_{HF}$–V curve with respect to that of an ideal structure, in flat band condition (see Fig. 3.2), by a quantity ΔV_{FB2} given by,

$$\Delta V_{FB2} = -\frac{Q_{it}t_{ox}}{\varepsilon_o \varepsilon_{ox}}. \tag{3.1}$$

where t_{ox} is the oxide thickness, ε_o the permittivity of the free space, ε_{ox} the relative permittivity of oxide.

However, at low frequency, the presence of interface-trapped charges modifies both the value of overall low frequency capacitance (whose minimum gets closer to C_{ox}) and relationship between C and V_G. This is because the interface-trapped charges have time to adjust their charge states. The capacitance of the structure is given by

$$\frac{1}{C} = \frac{1}{C_{ox}} + \frac{1}{C_D + C_{it}} \tag{3.2}$$

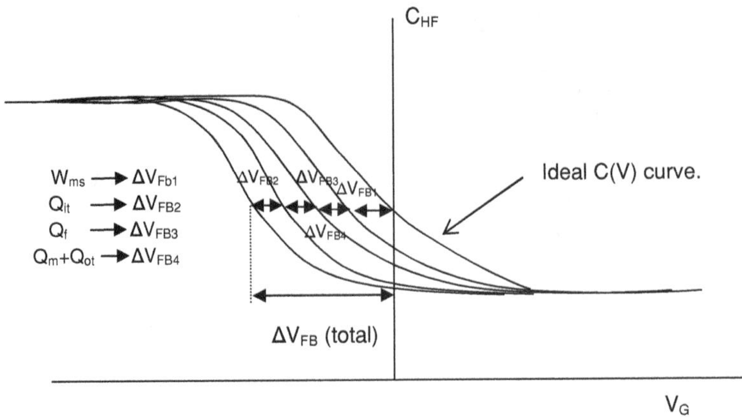

Fig. 3.2 Representation of the C–V curves of a P-type MOS structure showing the shifts introduced by work function difference and various oxide charges

where C_{it} represents the capacitance associated with the interface trapped charges and C_D represents the depletion capacitance.

Several techniques have been used to measure the density and density-distribution of the interface trapped charges in MOS device [8]. Most of these techniques are based on the measurement of MOS capacitance such as capacitance–voltage (C–V) method [19–21], deep-level-transient-spectroscopy (DLTS) [22–24], and photo-capacitance-transient spectroscopy (PCTS) [25]. Some of them utilize low-frequency-capacitance e.g. C–V method, whereas others utilize high-frequency-capacitance e.g. DLTS and PCTS. The C–V method of analysis has helped to provide qualitative or even semi-quantitative measurements in the range of trap-density 10^{10} or 10^{11} cm^{-2}. Comparatively, a few numbers of methods have been developed which make use of the measurement of other electric quantities for studying the interface-trapped charges. For example, in conductance method [17] the frequency dependent conductance of a MOS device is used to extract information about its surface states. In charge pumping-method [26–30], measurement of the dc substrate-current is utilized for studying the interface-trapped charges. However, the use of the charge-pumping technique is limited only to certain specified MOSFETs. It is well known that the interface-trapped charge can be neutralized by low-temperature (450°C) hydrogen annealing. Furthermore, annealing MOS structure in chlorine ambient can reduce its density [31].

3.6.2 The Fixed Oxide Charge

Fixed oxide charge is positive charge, primarily due to structural defects (oxidation process) in the oxide [32, 33]. It is located in the oxide within 25 Å of the silicon surface. This fixed oxide charge, Q_f, differs greatly from the interface trapped charge in that its density magnitude. For all practical purposes, it is not a function of the applied gate voltage or the surface potential in the silicon near the interface because the energy levels of the states associated with this type of charge lie outside the forbidden band gap. Its density is constant even under thermal or electrical stressing, which would normally cause movement of mobile ionic charges. Thus, it is not in any way related to sodium and other ions of mobile charge contamination that might be introduced during device fabrication. The type or concentration of doping impurities in the silicon does not significantly affect by the oxide thickness or the density of fixed oxide charge. But, since its origin is related to the oxidation process, then it depends on oxidation ambient and temperature, cooling conditions, and on silicon crystal orientation.

The observed density of the fixed charge N_f as a function of silicon substrate orientation under similar processing conditions decreases in the order $(111) > (110) > (100)$ in approximately the ratio 3:2:1. Typical values of the density of fixed oxide charge, for an optimized fabrication process are below 10^{10} cm^{-2}.

These charges have an impact similar to other oxide charges on the C–V curve. They cause a shift ΔV_{FB3} of the C(V) curve with respect to that of an ideal structure as illustrated in Fig. 3.2 and given by

$$\Delta V_{FB3} = \frac{Q_f t_{ox}}{\varepsilon_o \varepsilon_{ox}} \tag{3.3}$$

In the above formula the fixed oxide charge is considered as a charge sheet located at the oxide–silicon interface. The fixed oxide charge density can be determined using the high-frequency capacitance–voltage (C–V) technique [34]. Since the density of this type of charges can not be determined unambiguously in the presence of moderate densities of interface trapped charge, it is only measured after a low temperature (450°C) hydrogen treatment which minimizes interface trapped charge density.

3.6.3 The Oxide Trapped Charge

Oxide trapped charges (Q_{ot}) may be positive or negative due to holes or electrons trapped in the bulk of the oxide, similar to interface trapped charges with the only difference that they exist in the bulk [35]. Trapping may be resulted from any phenomenon which either create or inject carriers in the bulk of the oxide such as ionizing radiation (including X-ray, gamma ray, low and high-energy electron irradiation) or avalanche injection. Like the interface-trapped charges, a low temperature (300°C) could eliminate the oxide-trapped charges induced by radiation, anneal in an inert ambient. Either the oxide trapped charge distribution or the total trapped charge or its centroid can be quantitatively determined in the oxide by using the most common methods such as etch-off C–V method [36] or photo I–V method [37–38]. The former method can be investigated by etching off the oxide thickness and measuring either the number of traps in etched-off oxide or the quantity of traps remaining in the oxide by C–V technique. The latter method was used by DiMaria et al. [35] for determining the total oxide trapped charge and the charge centroid. Przewlocki [38] also used this method to determine the trapped charge distribution.

If the oxide charges contained in the oxide are only due to oxide trapped charges Q_{ot} a change in the distribution, caused by an excitation (temperature, photons), gives rise to a new value of V_{FB} of Eq. (2.15).

$$\Delta V_{FB4} = - \int_0^{t_{ox}} \frac{\rho(x)x\,\mathrm{d}x}{\varepsilon_0 \varepsilon_{ox}} \tag{3.4}$$

This effect can be used to calculate the total amount of trapped charge in the oxide.

3.6.4 The Mobile Ionic Charge

A mobile ionic charge in thermal oxides were the first charge to be extensively investigated since further studies in the SiO_2 system could not be carried out until the mobile ions level was minimized. The mobile ionic charge is due to ionic impurities such as Li^+, Na^+, K^+, and possibly H^+. This charge can easily move from one edge to the other of the oxide layer under thermal-electrical stress, and the resulting movement of charge can cause an unwanted instability and a change in the electrical device parameters [8, 39–42]. The complete study of the mobile ionic charge effects and its density determination are discussed in details in the next chapters. The effect of ionic drift caused by electric field can be effectively minimized in a number of ways. The most commonly employed approach is to eliminate, with ultra-clean processing techniques, as much ions to contamination in the oxide as possible. For example, great care must be taken to keep the quartz walls of the furnace in which the oxide is grown virtually sodium free. Kriegler et al. [43] have reported that a mixture of hydrogen chloride gas and dry oxygen is extremely effective for the "cleaning" of quartz furnace tubes. They also found that the addition of a small percentage of hydrogen chloride or chlorine to the oxidizing atmosphere significantly improved the electrical stability of SiO_2 films grown in the presence of dry oxygen. They found that this technique not only decreases the mobile ion contamination originating from the furnace tube, but also tends to passivate the oxide films grown in this manner against ionic instabilities caused by the subsequent deposition of a contaminated metal layer. By using radioactive tracers, Yon et al. [44] showed that sodium ions tend to be much more soluble in the phospho-silicate glass than in the oxide below and consequently are guttered by the glass layer. Thus preventing them from drifting across the oxide under the influence of the applied gate voltage.

If the charges contained in the oxide are only due to mobile ions Q_m a change in the distribution, caused by an applied stress (temperature + electric field), produces also a shift in the V_{FB} by an amount ΔV_{FB4} using Eq. (3.4). This effect can be used to calculate the total amount or the density of mobile ions. However, if both mobile ions and trapped charges are present, their redistributions will cause a same voltage shift ΔV_{FB4} under BTS and hence their effects cannot be separated.

References

1. Deal, B.E.: Standardized terminology for oxide charge associated with thermally oxidized silicon. IEEE Trans. Elect. Dev. **ED-27**, 606–608 (1980)
2. Schlegel, E., Schnable, G., Schwartz, R., Spratt, J.: Behavior of surface ions on semiconductor devices. IEEE Trans. Electron. Devices **ED-15**, 973–979 (1968)
3. McPherson, J.: Accelerated Testing. Electronic Materials Handbook. ASM International Publishing, Materials Park, OH (1989)
4. Snow, E.H., Grove, A.S., Deal, B.E., et al.: Ion transport phenomena in insulating films. J. Appl. Phys. **36**, 1664–1673 (1965)
5. Snow, E.H., Deal, B.E.: Polarization phenomena and other properties of phosphosilicate glass films on silicon. J. Electrochem. Soc. **113**, 263 (1966)
6. Hefley, P.L., McPherson, J.: The impact of an external sodium diffusion source on the reliability of MOS circuitry. IEEE-IRPS Proceedings, pp. 167–172 (1988)
7. Stuart, D.A.: Calculations of activation energy of ionic conductivity in silica glass by classical methods. J. Am. Ceram. Soc. 573–580 (1954)
8. Nicollian, E.H., Brews, J.R.: MOS Physics and Technology. Wiley, New York (1982)
9. Terman, L.M.: An investigation of surface states at silicon–silicon oxide interface employing metal oxide silicon diodes. Solid-State Electron. **5**, 285–299 (1962)
10. McNutt, M.J., Sah, C.T.: Determination of the MOS oxide capacitance. J. Appl. Phys. **46**, 3909–3913 (1975)
11. Mitra, V., Bentarzi, H., Bouderbala, R., et al.: A theoretical model for the density-distribution of mobile ions in the oxide of the metal-oxide-semiconductor structures. J. Appl. Phys. **73**, 4287–4291 (1993)
12. DiMaria, D.J.: Defects and impurities in thermal SiO_2. In: Partelides, S.T. (ed.) The Physics of SiO_2 and its Interface, pp. 160–178. Pergamon, New York (1978)
13. Sungano, T.: Recent understanding of morphology of Si–SiO_2 interface and traps states at the interface. In: Jain, S.C., Radhakrishna, S. (eds.) Physics of Semiconductor Devices, Wiley, Singapore (1982)
14. Ash, M.C., Chattopadhyay, P., Daw, A.N.: Effect of trichloroethylene on the oxide charge and interface state density of a silicon MIS tunnel structure. J. Inst. Electron. Telecommun. Eng. **31**, 63–64 (1985)
15. Avniand, E., Shapir, J.: Modeling of charge injection effects in metal-oxide-semiconductor structures. J. Appl. Phys. **64**, 734–742 (1988)
16. Litovchenko, V.G., Kiblick, V.Y., Georgiev, S.S., et al.: Radiation induced charges in low-temperature oxide MOS structures (Al–SiO_2–Si). Radiat. Eff. **62**, 1–5 (1982)
17. Nicollian, E.H., Goetzberger, A.: The Si–SiO_2 interface: electrical properties as determined by the MIS conductance technique. Bell Syst. Tech. J. **46**, 1055–1133 (1967)
18. Brown, D.M., Gray, P.V.: Si–SiO_2 fast interface state measurements. J. Electrochem. Soc. **115**, 760–766 (1968)
19. Gourrier, S., Friedel, P.: Caracterisation electronique des etats electroniques d'interface Isolant/Semiconductor. Acta Electron. **25**, 217–240 (1983)

20. Berglund, C.N.: Surface states of steam-grown silicon dioxide interfaces. IEEE Trans. Elect. Dev. **ED-13**, 701–705 (1966)
21. Kuhn, M.: A quasi-static technique for MOS C-V and surface state measurements. Solid State Electron. **13**, 873–885 (1970)
22. Lang, D.V.: Deep-level transient spectroscopy: a method to characterize traps in semiconductor. J. Appl. Phys. **45**, 3023–3032 (1974)
23. Wang, K.L., Evwaraye, A.O.: Determination of interface and bulk trap states of IGFET's using deep-level transient spectroscopy. J. Appl. Phys. **47**, 4574–4577 (1976)
24. Yamasaki, K., Sugano, T.: Determination of the interface states in GaAs MOS diodes by deep-level transient spectroscopy. Appl. Phys. Lett. **35**, 932–934 (1979)
25. Hasegawa, H., Sawada, T.: Proceedings of the 15th International Conference Physics of Semiconductors. Suppl. A **49**, 1125 (1980)
26. Brugler, J.S., Jespers, P.G.A.: Charge pumping in MOS devices. IEEE Trans. Electron. Devices **ED-16**, 297–301 (1969)
27. Elliot, A.B.M.: The use of charge pumping currents to measure surface state densities in MOS transistors. Solid State Electron. **19**, 241–247 (1976)
28. Groeseneken, G., Maes, H.E., Beltran, N., et al.: A reliable approach to charge pumping measurements in MOS-transistors. IEEE Trans. Elect. Dev. **ED-31**, 42–53 (1984)
29. Tseng, W.L.: A new charge pumping method of measuring Si–SiO$_2$ interface states. J. Appl. Phys. **62**, 591–599 (1987)
30. Heremans, P., Witters, J., Groeseneken, G., et al.: Analysis of the charge pumping technique and its application for the evaluation of MOSFET degradation. IEEE Trans. Elect. Dev. **ED-36**, 1318–1335 (1989)
31. Ho, V.Q., Sugano, T.: An temperature of the interface properties of plasma anodized SiO$_2$/Si system for the fabrication of MOSFET's. IEEE Trans. Elect. Dev. **ED-28**, 1060–1064 (1981)
32. Learn, A.J., Hess, D.W.: Effects of ion implantation on charges in the silicon–silicon dioxide system. J. Appl. Phys. **48**, 308–312 (1977)
33. Deal, B.E., Sklar, M., Grove, A.S., et al.: Characteristics of the thermally oxidized silicon. J. Electrochem. Soc. **114**, 226–274 (1967)
34. Tang, S.M., Berry, W.B., Kwor, R., et al.: High frequency capacitance-voltage characteristics of thermally grown SiO$_2$ films on β-Sic. J. Electrochem. Soc. **137**, 221–225 (1990)
35. DiMaria, D.J.: The properties of electron and holes traps in thermal silicon dioxide layers grown on silicon. In: Partelides, S.T. (ed.) The Physics of Si–SiO$_2$ and Its Interface, pp. 160–178. Pergamon, New York (1978)
36. Berglund, C.N., Powell, R.J.: Photoinjection into SiO$_2$: electron scattering in the image force potential well. J. Appl. Phys. **42**, 573–579 (1971)
37. Brews, J.R.: Limitations upon photo-injection studies of charge distributions close to interfaces in MOS capacitors. J. Appl. Phys. **44**, 379–384 (1973)
38. Przewlocki, H.M.: Determination of trapped charge distributions in the dielectric of a metal-oxide-semiconductor structure. J. Appl. Phys. **57**, 5359–5366 (1985)
39. Sze, S.M.: Physics of Semiconductor Devices. Wiley, New York (1981)
40. Grove, A.S.: Physics and Technology of Semiconductor Devices. Wiley, New York (1967)
41. Davis, J.R.: Instabilities in MOS Devices. Gordon and Breach Science Publishers, London (1977)
42. Hillen, M.W., Verwey, J.F.: Mobile ions in SiO$_2$ layers on Si. In: Barbottain, G., Vapaille, A. (eds.) Instabilities in Silicon Devices, pp. 404–439. North-Holland, Amsterdam (1986)
43. Kriegler, Y., Cheng, C., Colton, D.R.: The effect of HCl and Cl$_2$ on the thermal oxidation of silicon. J. Electrochem. Soc. **119**, 388–392 (1972)
44. Yon, E., Ko, W.H., Kuper, A.B.: Sodium distribution in thermal oxide on silicon by radiochemical and MOS analysis. IEEE Trans. Elect. Dev. **ED-13**, 276–280 (1966)

Chapter 4
Review of Transport Mechanism in Thin Oxides of MOS Devices

4.1 Introduction

There has been intensive research on the outermost limit of the downscaling of silicon dioxide films. A decrease in device size and the oxide thickness leads to a relative increase in the gate current. This current can have a significant effect on the MOS transistor operation. It can result from the flow of either *carriers* (electronic current) or *ions* (ionic current) or from the both. The different conduction mechanisms can be described under the condition that the oxide layer is wholly homogeneous. This condition can be attained for layers thicker than 50 Å but rarely attained for thin layers. When the oxide is not homogeneous, it is difficult to take in consideration geometrical parameters in the physical models.

4.2 Electronic Conduction

Measured gate current in thin oxides layer may be due to different mechanisms of electronic conduction. These carriers may be *intrinsic* or *extrinsic* (injected from the gate or the substrate). In the former, the conduction is of the ohmic type with high resistivity. The Ohmic current is always referred as leakage current. In the latter, the electrons may freely travel inside the oxide layer (Schottky effect, direct tunnel and Fowler–Nordheim tunnel effect), or their transport may be associated with traps (Frenkel-Poole effect, hopping conduction, and space charge limited current). The different types of electronic conduction for NMOS device are illustrated in Fig. 4.1 [1].

H. Bentarzi, *Transport in Metal-Oxide-Semiconductor Structures*,
Engineering Materials, DOI: 10.1007/978-3-642-16304-3_4,

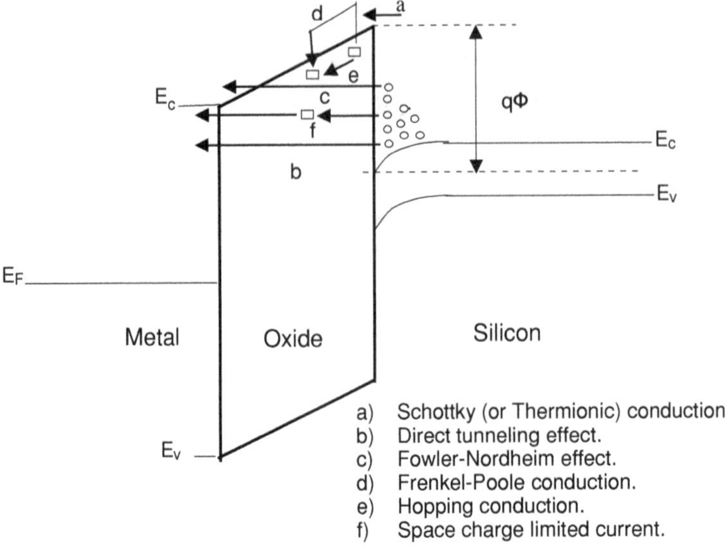

Fig. 4.1 Different types of electronic conduction in the oxide of NMOS Transistor

4.2.1 The Schottky (or Thermionic) Conduction

Schottky effect can appear if the energy of the electrons is sufficient to launch them into the oxide conduction band. In other words, the Schottky current is due to the electrons that move above the potential barrier, i.e. those with energy $E > q\phi$. This current can be expressed as [2].

$$J = \frac{4\pi m^* q k^2 T^2}{h^3} \exp\left(-\frac{q\phi}{kT}\right) \exp\left(\frac{q}{kT}\sqrt{\frac{q\zeta_{ox}}{4\pi\varepsilon_o\varepsilon_{ox}}}\right) \tag{4.1}$$

where m^* is the effective mass of electron, $q\phi$ the barrier height at the oxide-silicon interface, ζ_{ox} the electric field strength in the oxide, and h the Plank's constant.

In this case, the current flow is due to a thermally activated process. The electrons must acquire sufficient energy to jump over the barrier and cross the oxide.

4.2.2 The Tunneling Conduction

The tunnel emission is caused by field ionization of trapped electrons into the conduction band or by electrons tunneling from the metal Fermi energy level into the oxide conduction band. The computation of tunneling current necessitates knowing the number of electrons suspected to travel by tunnel effect, the energy distribution of these electrons (given by Fermi–Dirac distribution function), and the transmission or tunneling probability of an electron whose energy E would cross the barrier [3].Then, the approximate general equation of tunneling current density is given by [4]:

$$J = \frac{4\pi q m_o}{h^3} \int_{E_C}^{E_{\max}} T(E) kT \ln\left(1 + \exp\left(-\frac{E - E_F}{kT}\right)\right) dE \qquad (4.2)$$

where $E_{\max} = q\phi$ and m_o is the free electron mass. For the direct tunnel conduction, the current density can be evaluated from Eq. (4.2). Then, its approximate formula can be expressed as [5, 6]

$$J = \frac{q^2 m_o \zeta_{ox}^2}{8\pi . h . (q\phi - E_F) m_{ox} \left[1 - \left(1 - \frac{q\zeta_{ox} t_{ox}}{q\phi - E_F}\right)^{1/2}\right]^2}$$
$$\times \exp\left(\frac{-4\sqrt{2m_{ox}}}{3\hbar q \zeta_{ox}}\left[(q\phi - E_F)^{3/2} - (q\phi - E_F - q\zeta_{ox} t_{ox})^{3/2}\right]\right) \qquad (4.3)$$

where, m_{ox} is the effective electron mass in the oxide. The tunnel emission has the strongest dependence on the applied voltage but is essentially independent of the temperature.

The direct tunneling current through the oxide becomes significant when the film thickness decreases to below 50 Å. This may cause problems in device operation and also during characterization (Fig. 4.2). For ultra-thin oxide the direct tunneling current increases by about one order of magnitude when the thickness of the silicon dioxide layer decreases by one nanometer [6].

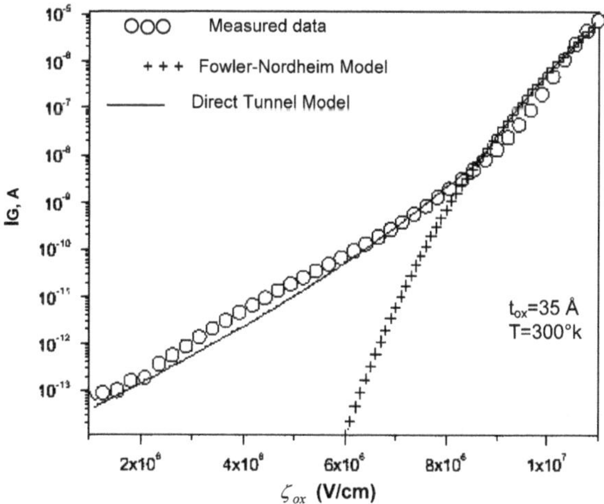

Fig. 4.2 The gate current modeling (using Direct tunnel and Fowler–Nordheim effect) of transistor NMOS with oxide thickness 35 Å (Reproduced with permission AIP [6])

4.2.3 The Fowler–Nordheim Conduction

The tunneling current density J, which describes the Fowler–Nordheim type, can be derived from Eq. (4.2). It is given by [7]:

$$J = \frac{q^2 m_{si}\zeta_{ox}^2}{8\pi h(q\Phi - E_F)m_{ox}} \exp\left(\frac{-4\sqrt{2m_{ox}}}{3\hbar q\zeta_{ox}}\left[(q\Phi - E_F)^{3/2}\right]\right) \qquad (4.4)$$

This mechanism is governed also by the vertical electric field ζ_{ox} across the gate oxide. It occurs, however, only at very high fields, larger than about 6 MV/cm [8]. Oxide damage due to Fowler–Nordheim tunneling is negligible for most MOSFET except for very thin oxide devices, as shown in earlier experimental results of spatial profiling of interface states [9]. It has been estimated [10] that direct tunneling dominates for oxides thinner than 40 Å and FN-tunneling for oxides over 50 Å. This phenomenon, however, has been used to determine the oxide thickness, as the oscillation period is dependent on the tunneling barrier thickness [10]. FN-current causes also some stress on the oxide which in turn changes oxide charges density that becomes unstable upon continuing electrical measurement [11]. This, further, causes instability, possibly making it more difficult to define the material parameters.

4.2.4 The Frenkel-Poole Conduction

The Frenkel-Poole emission is due to the field-enhanced thermal excitations of the trapped electrons into the conduction band. The expression of the Frenkel-Poole current for thin oxides is given by [11]:

$$J = qN_c\mu_e\zeta_{ox}\exp\left(-\frac{q\phi}{kT}\right)\exp\left(\sqrt{\frac{q}{\pi\varepsilon_o\varepsilon_{ox}}}\frac{q\sqrt{\zeta_{ox}}}{kT}\right). \qquad (4.5)$$

where N_c is the density of states in the oxide conduction band and μ_e is the mobility of electrons in the oxide.

4.2.5 The Hopping Conduction

In the case of hopping conduction, the energy of the electron is less than the maximum energy of height trap potential well. The hopping current can be derived, using the following equation [12]:

$$J = \frac{q^2}{kT}n^*\alpha^2\zeta_{\text{ox}}\frac{1}{\tau_o}\exp\left(\frac{-4\pi m^*}{h}\phi_m\alpha\right)$$
(4.6)

where n^* is the electron density on the sites, α is the distance between two sites, τ_o is the time constant and ϕ_m is the barrier height between two sites (the other symbols have their usual meanings).

4.2.6 The Space Charge-Limited Current

In the case of Frenkel-Poole and hopping conduction, the electric field is assumed constant. When the electron injection is strong, this hypothesis is no more valid and the potential distribution should be calculated using Poisson's equation. The space-charge-limited current results from a carrier injected into the oxide, where no compensating charge is present. This current has different expressions for different injection [13].

Weak Injection

When the injection is weak, the oxide charge density is negligible and the electric field is therefore constant. The current is given by

$$J = qn(x)\mu_e\frac{V}{t_{\text{ox}}}$$
(4.7)

where $n(x)$ is the number of conduction electrons.

Strong Injection

In this case, the oxide traps fill up and a space charge builds up. The current is given by

$$J = \frac{8}{9}\mu_e\frac{\varepsilon_o\varepsilon_{ox}}{t_{\text{ox}}^3}V^2\frac{N_c}{N_t}\exp\left(-\frac{q\phi_t}{kT}\right)$$
(4.8)

with N_t is the trap density, N_c is the density of states in conduction band, and $q\phi_t$ the energy difference between the conduction band and a trapping site.

Very Strong Injection

In this case, all traps being filled, the space charge is due to conduction electrons and the current can be expressed as:

$$J = \frac{8}{9}\mu_e\frac{\varepsilon_o\varepsilon_{ox}}{t_{ox}^3}V^2 \tag{4.9}$$

For current–voltage characteristics in accumulation mode with continuous electron injection from the gate, the space-charge-limited current is not considered a major contribution to the dc current through the capacitor.

At low voltage and high temperature, thermally excited electrons, hopping from one isolated state to another, carry current. This mechanism yields an Ohmic characteristic exponentially dependent on temperature.

4.3 Ionic Conduction

Ionic conduction is identical to the ohmic conduction except that the ionic current resulting from the ionic conduction decreases with temperature. The ionic conduction is similar to a diffusion process. Generally, when the electric field is applied, the dc ionic conductivity decreases with time, because ions cannot be readily injected into or extracted from the oxide. After an initial current flow, positive and negative space charges will build up near the metal-oxide and the semiconductor-oxide interfaces, causing a distortion of the potential distribution. When the applied field is removed, large internal fields remain which cause ions to flow back toward their equilibrium positions, thus resulting in a hysteresis effect. The J-V characteristic for ionic conduction is given by [12]:

$$J = B\zeta_{ox}T^{-1}\exp\left(\frac{-\Delta E_{ai}}{kT}\right) \tag{4.10}$$

where B is a constant, and ΔE_{ai} is the activation energy for ion migration. According to Eq. (4.10), log (JT) versus $1/T$ should yield a straight line with a slope determined by the activation energy for ion migration (the slope is independent of gate bias). The plots of log (JT) versus $1/T$ are linear as shown in Fig. 4.2, but the slopes change with gate bias, and J is not linearly dependent on V as required for ionic conduction. If there is ionic movement in a gate under bias, it will result in a flat-band voltage shift in the C–V data during bias-temperature stress (BTS) measurement [12]. The BTS test did not show any effect that can be attributed to a mobile charge; therefore, ion movement is not an issue.

Another factor that rules out an ionic conduction mechanism is that the conduction should be independent of barrier height, which means positive or negative bias should induce the same current as long as the electric field in the oxide is the same. All the I–V characteristics show that, for a negative bias in accumulation mode, the leakage current is higher than in the positive bias case (i.e., the inversion region).

4.3.1 Ionic Current Transport Equation

An applied electric field and a gradient in mobile ion concentration can cause current flow in the oxide layer. The first causes ion drift, and the second gives rise to ion diffusion.

Drift Current

Drift current is due to mobile ion acceleration is caused by an applied electric field. The current density is given by:

$$\vec{J}_{\text{drift}} = +qN_m\mu_{\text{ion}}\vec{\zeta} \tag{4.11}$$

where N_m, μ_{ion}, $\vec{\zeta}$ denote respectively the ion density, the mobility of ion and the electric field.

Diffusion Current

This type of current is caused by the gradient in ion concentration, in other words, ions diffuse from region of higher density to region of lower density. The density of the diffusion current is given by

$$\vec{J}_{\text{diffusion}} = -qD\nabla N_m. \tag{4.12}$$

where ∇N_m is the gradient of mobile ion density, and D is the ion diffusion constant that is expressed by the Einstein relationship: $D = \frac{\mu_{\text{ion}}kT}{q}$.

When both concentration gradient and electric field present, the total ionic current density is

$$\vec{J}_{ion} = qN_m\mu_{\text{ion}}\vec{\zeta} - qD\nabla N_m \tag{4.13}$$

However, when high density of ion is present, the ionic current density can be derived from Eq. (4.13), which is given by

$$J = q.N_m.\mu_{\text{ion}}.\zeta_{\text{ox}} - qD\frac{\partial N}{\partial x} \tag{4.14}$$

where $\frac{\partial N}{\partial x}$ is the gradient of ion density and D is the ion diffusion constant.

4.4 Summary

For a given oxide thickness, each conduction process type may dominate in a certain range of temperature and bias. The tunnel emission has the strongest

Table 4.1 Basic conduction mechanisms and their temperature and voltage dependence characteristics

Process	Expression	Temperature and voltage dependence
Schottky emission	$J = \frac{4\pi m^{*} q k^{2} T^{2}}{h^{3}} \exp\left(-\frac{q\phi}{kT}\right) \exp\left(\frac{q}{kT}\sqrt{\frac{q\zeta_{\mathrm{ox}}}{4\pi\varepsilon_{o}\varepsilon_{\mathrm{ox}}}}\right)$	$\propto T^{2} \exp\left(a.\frac{V^{1/2}}{T}\right)$
Direct tunneling	$J \propto .\zeta_{\mathrm{ox}}^{2} \exp\left(\frac{-4\sqrt{2m_{\mathrm{ox}}}}{3\hbar q\zeta_{\mathrm{ox}}}.\left[(q\phi - E_{F})^{3/2} - (q\phi - E_{F} - q\zeta_{\mathrm{ox}}t_{\mathrm{ox}})^{3/2}\right]\right)$	$\propto V \exp\left(-\frac{b}{V}\right)$
Fowler–Nordheim	$J \propto \zeta_{\mathrm{ox}}^{2} \exp\left(\frac{-4\sqrt{2m_{\mathrm{ox}}}}{3\hbar q\zeta_{\mathrm{ox}}}.\left[(q\phi - E_{\mathrm{F}})^{3/2}\right]\right)$	$\propto V \exp\left(-\frac{b}{V}\right)$
Frenkel–Poole	$J = q.N_{\mathrm{c}}.\mu_{e}.\zeta_{\mathrm{ox}}.\exp\left(-\frac{q\phi}{kT}\right).\exp\left(\sqrt{\frac{q}{\pi.\varepsilon_{o}\varepsilon_{ox}}}\frac{q\sqrt{\zeta_{\mathrm{ox}}}}{kT}.\right).$	$\propto V \exp\left(2.a.\frac{V^{1/2}}{T}\right)$
Space-charge-limited	$J = \frac{8}{9}.\mu_{e}.\frac{\varepsilon_{o}\varepsilon_{ox}}{t_{\mathrm{ox}}^{3}}.V^{2}$	$\propto V^{2}$
Ohmic conduction	$J = A.\zeta_{\mathrm{ox}}.\exp\left(\frac{-\Delta E_{\mathrm{ae}}}{kT}\right)$	$\propto V \exp\left(-\frac{c}{T}\right)$
Ionic conduction	$J = B.\zeta_{\mathrm{ox}}.T^{-1}.\exp\left(\frac{-\Delta E_{\mathrm{ai}}}{kT}\right)$	$\propto \frac{V}{T}\exp\left(-\frac{d}{V}\right)$

A and B are constants, ΔE_{ae} = activation energy of electrons, ΔE_{ai} = activation energy of ions. $V = \zeta_{\mathrm{ox}}t_{\mathrm{ox}}$. Positive constants independent of V or T are b, c, and d (The other symbols have their usual meanings)

dependence on the applied voltage and the Schottky emission has the strongest dependence on the temperature. The space charge limited process is dependent on the applied gate voltage. The Frenkel-Poole process is important at moderate temperatures and fields. The Ohmic and ionic conduction processes are of less importance among the above-described phenomena, however; the ionic process is severely affected by high temperature. The hopping process shows the same dependence on temperature and voltage as the ionic conduction. These processes are not exactly independent of one another and should be carefully examined. For example, for the large space charge effect, the tunneling characteristic is found to be similar to the Schottky type emission.

For an ultra-thin SiO_{2} gate dielectric, the dominating conduction mechanism is tunneling, while for alternative gate dielectrics that have a thicker physical thickness, the conduction mechanism will be different [1]. Current–voltage measurements at various temperatures were used to study the conduction mechanisms, which are based on the voltage and temperature dependence of the leakage current characteristics. Basic conduction mechanisms can be defined as summarized in Table 4.1.

References

1. Hesto, P.: Nature of electronic conduction. In: Barbottin, G., Vapaille, A. (eds.) Instabilities in silicon devices, vol. 1. North Holland, Amsterdam (1986)

2. Cohen-Tanoudji, C., Diu, B., Laloë, F.: Mécanique quantique Collection enseignement des sciences. Hermann éditeurs des sciences et des arts, Paris (2001)
3. Pananakakis, G., Ghibaudo, G., Kiès, R.: Temperature dependence of the Fowler-Nordheim current in metal-oxide-degenerate semiconductor structures. J. Appl. Phys. **78**, 2635 (1995)
4. Chang, C.: Tunneling in thin gate oxide MOS structures. Phd Thesis, Berkley, USA (1984)
5. Depas, M., Vermeire, B., Mertens, P.W., et al.: Determination of tunneling parameters in ultra-thin oxide layers poly-Si/SiO$_2$/Si structures. Solid State Electron. **38**, 1465–1471 (1995)
6. Maserjian, J., Zamani, N.: Behavior of the Si/SiO$_2$ interface observed by Fowler-Nordheim tunneling. J. Appl. Phys. **53**, 559–567 (1982)
7. Weinberg, Z.A.: On tunneling in metal-oxide-silicon structures. J. Appl. Phys. **53**, 5052–5056 (1982)
8. Li, X.M., Deen, M.J.: Determination of interface state density in MOSFETs using the spatial profiling charge pumping technique. Solid State Electron **35**, 1059–1063 (1992)
9. Hiroshima, M., Yasaka, T., Miyazaki, S., Hirose, M.: Electron tunneling through ultra thin gate oxide formed on hydrogen-terminated Si(lOO) surfaces. Jpn. J. Appl. Phys. **33**, 395–398 (1994)
10. Zafar, S., Liu, Q., Irene, E.A.: Roughness at the Si/SiO$_2$ interface. J. Vac. Sci. Technol. A **13**, 47 (1995)
11. Scarpa, A., Paccagnella, A., Ghidini, G. et al.: Instability of post-Fowler–Nordheim stress measurements of MOS devices. Solid State Electron. **41**, 935–938 (1997)
12. Schroder, D.K.: Semiconductor material and device characterization. Wiley, New York (1998)
13. Sze, S.M.: Physics of semiconductor devices. Wiley, New York (1981)

Chapter 5
Experimental Techniques

5.1 Introduction

Three standard techniques are commonly used to study the nature and the properties of these ions as well as to measure their density distributions [1–6]. These measurement techniques are:

- High Frequency MOS C–V measurement under Bias Thermal Stress (BTS), which is based on the observation of the shifts of MOS C–V characteristic under the influence of an applied gate voltage and temperature [1–3, 7],
- Triangular Voltage Sweep (TVS) method which utilizes the measurement of ion current producing from the applied of triangular voltage sweep to the gate [1, 8–12],
- Thermally Stimulated Ionic Current (TSIC) method, which uses the measurement of ionic current producing from the application of temperature to the gate [13–16].

Besides, we have recently developed a new method where charge-pumping technique may be used to measure the density of the mobile ions. This Charge-pumping current measurement is investigated before and after BTS. The obtained charge-pumping current curve can be used to extract the flat-band voltage shift that is produced by BTS stress. This flat-band voltage shift may be due to redistribution of the mobile ions as well as the trapped charges. This effect can be used to calculate the total amount of mobile ions or the density distribution of mobile ions. However, this can be done only if the effect due to a change in the trapped charge density in the oxide can be separated. Since this type of oxide charge will contribute in producing the charge pumping current then, the latter may be used to calculate the change in the trapped charge density.

The characterization tools used to investigate these techniques are shown in Fig. 5.1. These measuring instruments are arranged and controlled by PC in such a way that measured data and needed parameters are extracted with high resolution.

H. Bentarzi, *Transport in Metal-Oxide-Semiconductor Structures,*
Engineering Materials, DOI: 10.1007/978-3-642-16304-3_5,
© Springer-Verlag Berlin Heidelberg 2011

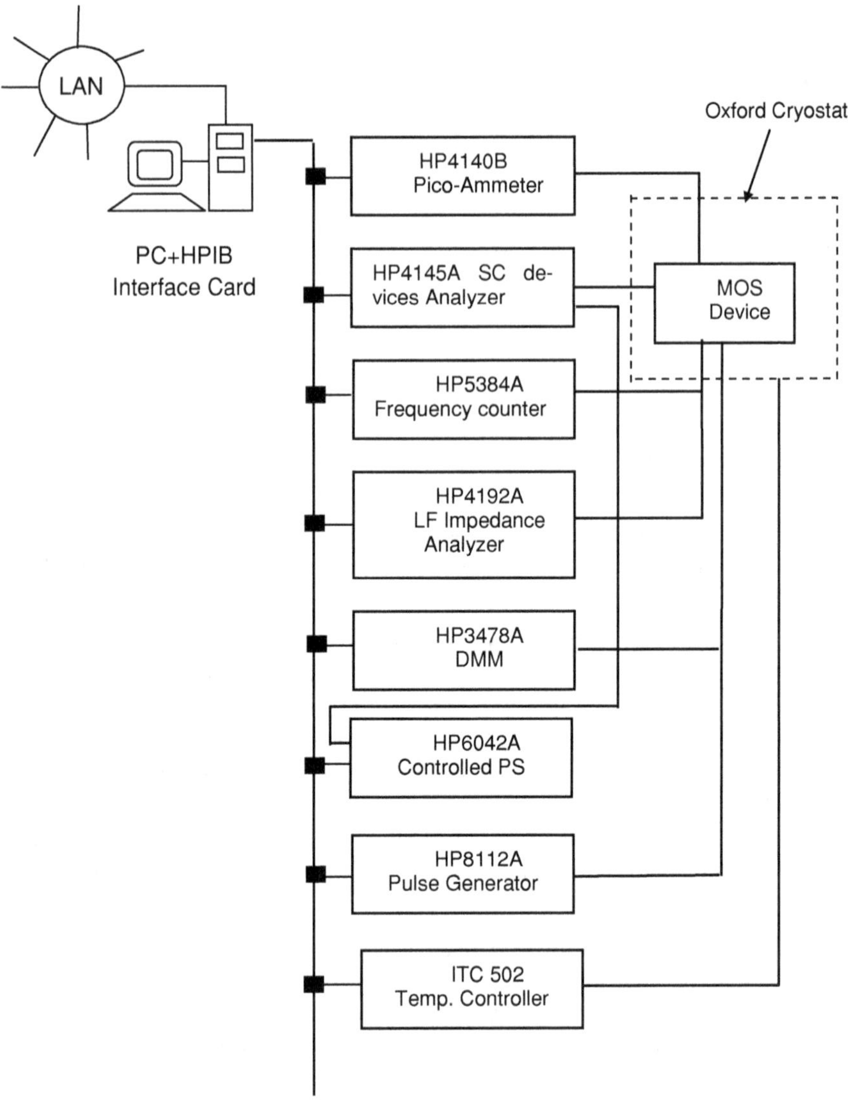

Fig. 5.1 General block diagram of the experimental set up system

Therefore, automated operation from measurement to analysis is necessary. This can be achieved by using instruments that can be remotely controlled by the computer. The IEEE bus (also known as, HPIB, GPIB, or ASCII bus) which is used as an interface bus for the Hewlett Packard instruments is one of the most popular data buses used in the industries and laboratories. It is a interfacing means that simplifies the integration of measuring HP instruments and the computer into a system [17].

5.2 High Frequency MOS C–V Measurement under BTS

In this experimental technique, the usual high frequencyC–V measurements under
bias-temperature stress are used to determine the mobile ion concentration within
the oxide layer in MOS capacitor. The high frequency (usually 1 MHz) C–V curve
of the given MOS capacitor is first measured before applying any stress. Then, the
MOS capacitor is heated to a certain temperature up to 250°C and held there for a
period up to 30 min, which is long enough to ensure that all the available ions drift
completely across the oxide. At the same time a positive gate bias is applied which
is enough to produce an oxide electric field of a few MV/cm. After holding the
MOS capacitor at elevated temperature and high electric field for the required
period, it is cooled back to room temperature so that no further redistribution of
charge takes place during the second C–V curve measurement. The flat band
voltage shift between theC–V curve before and after BTS is a measure of the
mobile ion concentration drifted at the given temperature.

5.2.1 Determination of the Flat-Band Voltage

This method is carried out by measuring the shifts in the flat-band voltage V_{FB}
under the influence of the BTS. A more reliable estimate of V_{FB} is obtained from
the portion of high frequency C–V curve corresponding to the depletion. The
measurement of V_{FB} from the C–V curve, needs to be carried out at a flat-band
capacitance C_{FB}. As the name indicates, the flat-band capacitance C_{FB} is the
capacitance of the MOS structure when the energy bands are flat near the Si–SiO$_2$
interface and has the following relation with the oxide capacitance [2]:

$$C_{FB} = \frac{C_{ox} C_{FBS}}{C_{ox} + C_{FBS}} \tag{5.1}$$

where C_{FBS} is the silicon flat-band capacitance that is given by

$$C_{FBS} = \frac{A\,\varepsilon_o\,\varepsilon_{ox}}{L_D} \tag{5.2}$$

where A is the gate area, ε_s the relative permittivity and ε_o the permittivity of the
free space. The extrinsic Debye length L_D can be expressed in terms of the sub-
strate doping concentration N_A as follows:

$$L_D = \sqrt{\frac{kT}{q^2 N_A}} \tag{5.3}$$

where k is the Boltzmann constant, q the electron charge and T the temperature in
Kelvin. The gate voltage corresponding to the flat-band capacitance on the high
frequency C–V curve is the flat-band voltage.

5.2.2 How the Mobile Charges Effect can be Separated

In the case of the real structure, the shift in the flat-band voltage is due to the sum of the effects of oxide charges and the work function difference. This section presents how to distinguish between flat-band voltage shift due to mobile ionic charge and those due to the other types of oxide charge. Considering an experiment where the only oxide charge is oxide fixed charge (Q_f), the initial high frequency C–V curve is labeled (0) in Fig. 5.2a. After heating at 180°C during a half hour with a positive gate bias (with respect to substrate) producing an electric field of a few million volts per centimeter across the oxide, and cooling back to room temperature, the curve labeled (+) in Fig. 5.2a is obtained. Repeating the bias-temperature stress with negative bias yields curve (−) in Fig. 5.2a. Figure 5.2a shows that no shift in the C–V curve is noticed. Therefore, the oxide fixed charge distribution does not change under this treatment. Repeating this bias temperature stress experiment with oxide trapped charge (Q_{ot}) that anneals out at low temperature, curves (0), (+), and (−) in Fig. 5.2b show that the oxide trapped charge density is changed. Gate bias polarity has no effect. Therefore, it is most likely that the oxide trapped charge

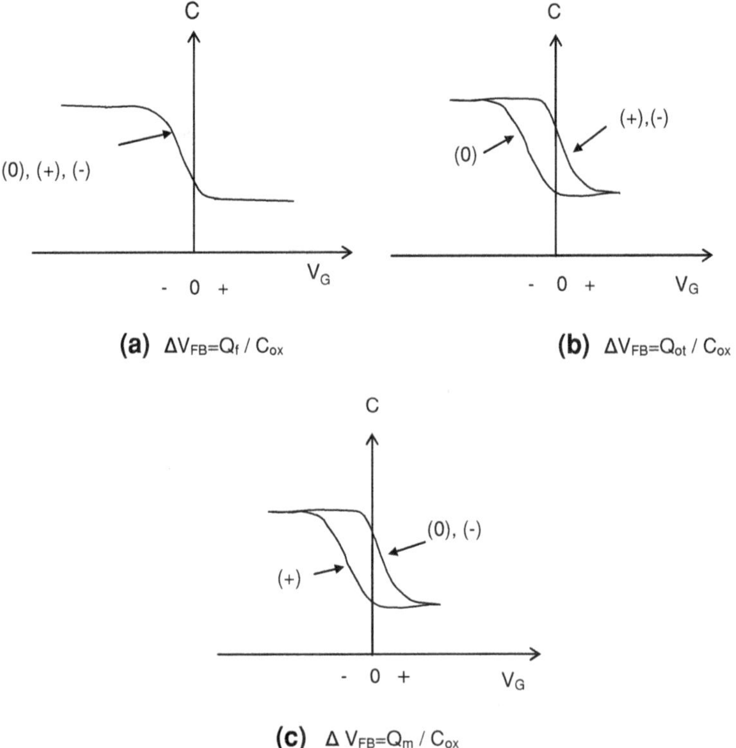

Fig. 5.2 Diagram illustration of ionic charge effect separation through the use of BTS method. (0) the initial CV curve, (+) after positive bias stress, (−) after negative bias stress

centers are immobile. Finally, Fig. 5.2c shows the results of repeating this experiment on an oxide contaminated by mobile ionic charge (Q_m). The flat band voltage V_{FBS} is initially low and after positive bias stress, it increases. With negative bias stress, V_{FB} returns to its original value. These results are due to mobile ion movement that affects on the flat-band voltage shift [2].

5.2.3 Theory

It is well known that the flat-band voltage of a MOS device undergoes a shift whenever there is any change in the concentration or redistribution of the mobile ions within its oxide-layer. A device fabricated under controlled conditions is supposed to have insignificant mobile ion-concentration and such a device will be referred as controlled device. If such a controlled device is intentionally contaminated further by introducing additional mobile ions, the flat-band voltage V_{FB} of the controlled device changes by an amount ΔV_{FB} given by [1]

$$\Delta V_{FB} = \frac{1}{\varepsilon_0 \varepsilon_{ox}} \int_0^{t_{ox}} x\rho(x)dx, \tag{5.4}$$

where $\rho(x)$ is the volume-density of the mobile ionic charge within the oxide and the distance x is measured from the metal-oxide interface. By assuming all the contaminated mobile ions to be concentrated in a thin charge sheet at either the metal-oxide or the silicon-oxide interface, Eq. (5.4) can be used to calculate charge density from the flat band voltage shift after introducing an average distance $\overline{X}$ called the centroid of the charge distribution. The charge centroid is defined by

$$\overline{X} = \frac{\int_0^{t_{ox}} x\rho(x)dx}{\int_0^{t_{ox}} \rho(x)dx}. \tag{5.5}$$

In addition, the total quantity of ions Q_{tot} is supposed to remain constant within the oxide before and after the drift and is given by

$$Q_{tot} = \int_0^{t_{ox}} \rho(x)dx \tag{5.6a}$$

with the help of Eq. (5.6a), Eq. (5.5) becomes

$$\Delta V_{FB} = -\frac{Q_{tot}}{C_{ox}} \tag{5.6b}$$

from Eqs. (5.4) and (5.6b), the flat band-voltage shift ΔV_{FB} of Eq. (5.4) can be rewritten

$$\Delta V_{FB} = \frac{\overline{X} Q_{tot}}{\varepsilon_o \varepsilon_{ox}} \tag{5.7}$$

When mobile ions can be considered as sheet of charge located at the Si–SiO$_2$ interface, Q_{tot} is obtained directly from a measurement of V_{FB} because the centroid is taken to be equal the oxide thickness. For this case, Eq. (5.7) becomes

$$\Delta V_{FB} = -\frac{t_{ox} Q_{tot}}{\varepsilon_o \varepsilon_{ox}} \tag{5.8a}$$

or

$$\Delta V_{FB} = -\frac{Q_{tot}}{C_{ox}} \tag{5.8b}$$

5.2.4 Experimental Results and Discussion

In BTS experiments, typical values are from 100°C to 250°C for a stress temperature, of the order of 20–30 min for a stress time, and in the range of 1–2 MV/cm for an electric field in the oxide caused by a stress voltage.

The experimental set up for measuring the capacitance is illustrated in Fig. 5.1. A gate voltage signal of constant amplitude that can sweep the energy level from the deep accumulation to the deep inversion is applied to a MOSFET. The drain and source are left open circuit. LCR meter (HP4140B) is connected to a substrate to measure the capacitance as function of gate voltage. Experimental measurements have been carried out on a number of commercial MOSFETs, such as n-MOSFET 3N171 and p-MOSFET 3N163, and encapsulated wafers in 64-pin PGA package with oxide thickness 40 and 12 nm respectively.

The flat band voltage shifts that are due to BTS stress were extracted from a measured high frequency CV curve as shown in Figs. 5.3 and 5.4. This shift can be used to determine the total amount of mobile ions in the oxide.

The results of mobile ion density as obtained by the measurement of the flat-band voltage shift ΔV_{FB} under bias-temperature ion drift are quite reliable and are not affected by any error in the measurement of V_{FB}, which may be caused by the trap level density at Si–SiO$_2$ interface in thick oxides. It is so because in high frequency C–V curves the trap level density is not supposed to undergo any significant change and any error in the measurement of V_{FB} due to trap levels, remaining constant before and after the application of bias temperature, cancels out in the determination of ΔV_{FB}. However, in thin film as the bias-temperature treatment causes small changes in interface trap level density, the C–V curves before and after bias-temperature stress are not the same but they are parallel to each other. Then, the flat band voltage shift is simply the parallel voltage shift. It may be noted that the shift in flat band voltage has a lower value when the same

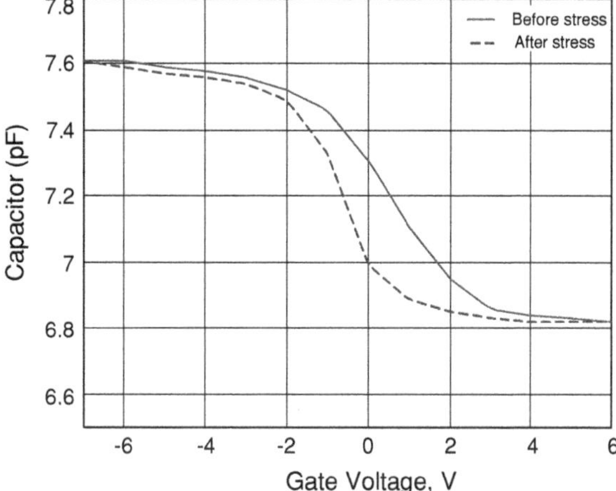

Fig. 5.3 High frequency capacitance-gate voltage curve of nMOSFET

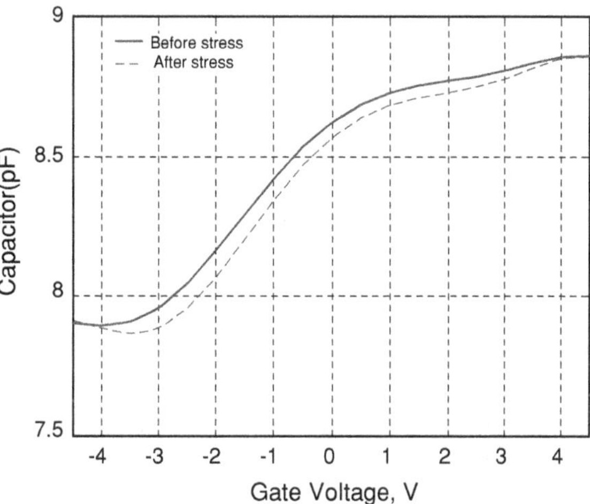

Fig. 5.4 High frequency capacitance as function of gate bias voltage for p-MOSFET

amount of ions is present in a thinner oxide. For current IC fabrication processes, a flat-band shift of 50 mV is acceptable which in a technology using a 100 nm oxide thickness corresponds to an ionic density of about 10^{10} ions cm^{-2}.

5.3 TVS Technique

A fast, simple and very sensitive technique is TVS method, which is capable of detecting better than 10^9 mobile ions/cm^2 and is based on the measurement of the displacement current response to a slow linear ramp voltage at elevated temperatures. This yields an ionic displacement current peak whose area is proportional

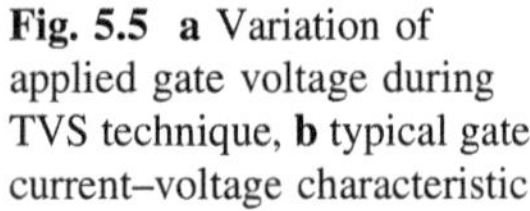

Fig. 5.5 a Variation of applied gate voltage during TVS technique, **b** typical gate current–voltage characteristic

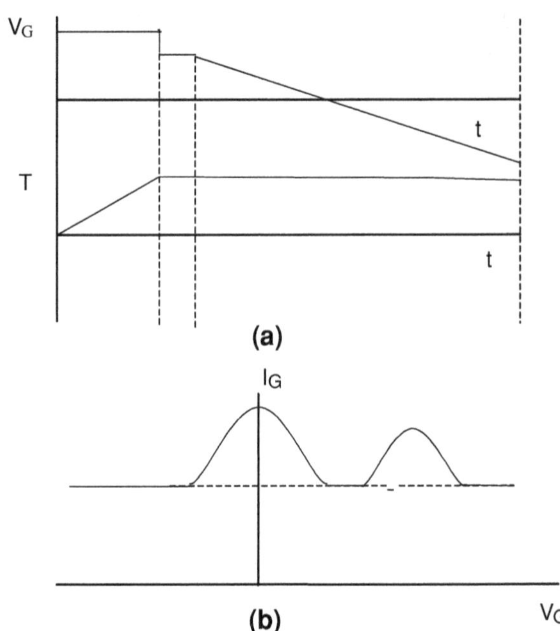

to the total mobile ionic charge. This method has been independently developed by Yamin [7] and Chou [8] who have tested and confirmed its validity using the simpler and electrochemically symmetrical Si(poly)–SiO_2–Si(100) MOS structure. This technique is expected to be very useful for routine process and quality control applications. Further more it has been used to study positive mobile charge behavior in the oxide [7, 9–11].

The experimental setup for performing mobile ion drift measurements using TVS technique is shown schematically in Fig. 5.1. Figure 5.5 shows the variation of applied voltages (gate voltage varies linearly with time).

In the TVS technique, the starting conditions are as follows:

- all mobile ions are initially at one of the interfaces,
- they are not trapped, but virtually free to move because the MOS capacitor is brought to elevated temperature.

The triangular ramp voltage which should be applied to the gate, is defined by

$$V_G = V_o + \alpha\, t \tag{5.9}$$

where t is time and $\alpha = dV/dt$ is sweep rate. A resulting gate current is measured against gate bias as the mobile ions drift from one interface to the other. Two modes can be distinguished: the quasistatic and non-quasistatic mode. In the former mode, the sweep rate α is chosen low enough during the entire measurement. The recording gate current is then similar to the quasistatic C–V curve, because

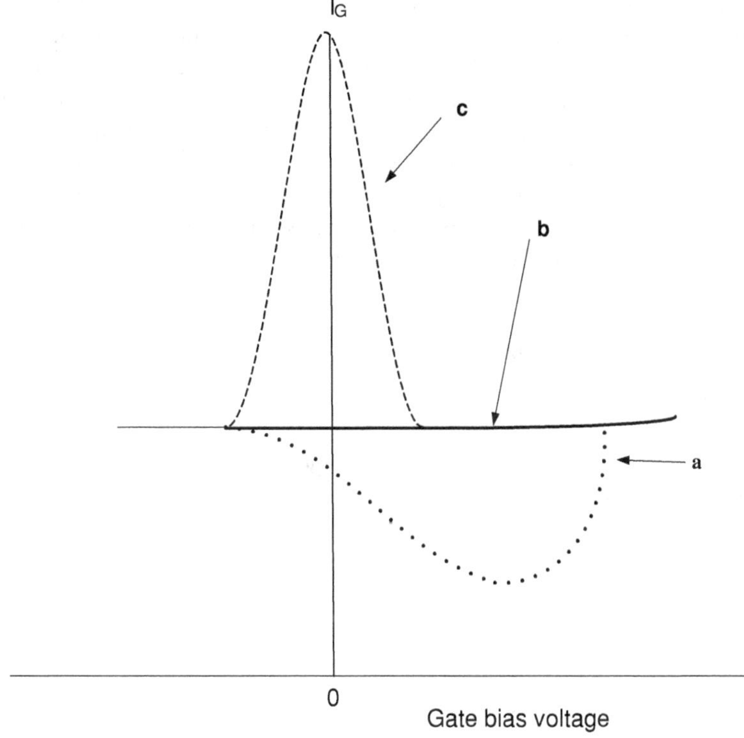

Fig. 5.6 Typical gate current response as function of applied voltage in a p-type MOS capacitor **a** at room temperature, **b** without mobile ions at 300°C, **c** with mobile ions at 300°C (Reproduced with permission, J. Electrochem. Soc. [9])

$$I_G = \frac{dQ_G}{dt} = \frac{dQ_G}{dV_G}\frac{dV_G}{dt} = \alpha\, C(V_G) \tag{5.10}$$

where $C(V_G)$ is the differential capacitance. However, in the latter mode, the sweep rate is high, and the ions can not follow the change in the applied electric field. In this situation, the TVS technique can be used either to determine accurately the number of ions that cross the oxide or to determine the mobility of the ions [17–19]. The idealized gate current response as function of applied bias with and without mobile ionic charge is shown in Fig. 5.6. At room temperature, the mobile ions are firmly trapped near the interfaces. In this situation, the observed I(V) curve is strictly similar to a quasistatic C(V) curve as shown in curve (a). At higher temperatures (typically about 300°C) and if no mobile ions are present, a quasi-static I(V) curve is obtained again, due to nearly constant value of capacitance equal to C_{ox} as shown in curve (b). If mobile ions are present, they contribute an additional component to the charging current of the MOS capacitor as shown in curve (c). The peak in the current of curve (c) arises as follows. At large negative gate bias all the mobile ions are at the metal-oxide interface and the gate current that flows is proportional to C_{ox}. As gate voltage increases, mobile ions begin drifting toward the Si–SiO_2 interface, attracting an increasing number of electrons to the silicon surface. That is, the ionic movement causes extra electrons to flow from the gate to the silicon through the external circuit, increasing thereby the gate current. This excess current peaks when the largest number of mobile ions is crossing the oxide layer. As gate voltage increases further, mobile ions pile up at the Si–SiO_2 interface. Consequently few number of ions flow, and the excess gate current falls. Ultimately, all the mobile ions that will drift at the given temperature have piled up at the Si–SiO_2 interface. At this stage, the gate current again becomes proportional to C_{ox}.

5.3.1 Theory

The mobile ion density drifted at a given temperature is proportional to the area under the peak in the gate current caused by the ionic motion. To show this, the ionic current per unit area I_G may be defined as

$$I_G = \frac{dQ_G}{dt} \tag{5.11}$$

where Q_G is the gate charge, which is given by

$$Q_G = C_{LF}(V_G - V_{FB}) \tag{5.12}$$

where $C_{LF}(V_G)$ is the low frequency MOS capacitance per unit area which is approximately equal to C_{ox} at elevated temperatures as shown in Fig. 5.6. Substitution of Eq. (5.10) in Eq. (5.9) gives,

$$I_G = C_{LF}(V_G)\frac{d(V_G - V_{FB})}{dt} \tag{5.13}$$

Equation (5.11) can be rewritten as

$$I_G = C_{ox}\left(\alpha - \frac{dV_{FB}}{dt}\right) \tag{5.14}$$

where α is the constant voltage sweep rate. Integrating from a gate bias $(-V_G)$ to a gate bias V_G, Eq. (5.12) gives

$$\begin{aligned}
\int_{-V_g}^{V_g} (I_G - \alpha\, C_{ox})dV_G &= -C_{ox}\int_{-V_g}^{V_g}\left(\frac{dV_{FB}}{dt}\right)dV_G, \\
&= -\alpha\, C_{ox}\int_{t(-V_G)}^{t(V_G)}\left(\frac{dV_{FB}}{dt}\right)dt, \\
&= -C_{ox}\{V_{FB}[t(V_G)] - V_{FB}[t(-V_G)]\}
\end{aligned} \tag{5.15}$$

The integration over gate bias on the right of Eq. (5.13) is carried out with respect to time. The integral on the left of Eq. (5.13) is the area between the $I_G(V_G)$ curve and the straight line $I_G = \alpha C_{ox}$, representing the gate current of the MOS capacitor when no ions move. The right side of Eq. (5.13) can be evaluated by using Eq. (5.7). Considering the mobile charge centroid to be located at $\overline{X}(-V_G)$ at time $t(-V_G)$ and $\overline{X}(V_G)$ at time $t(V_G)$, it can be shown that

$$V_{FB}[t(V_G)] - V_{FB}[t(-V_G)] = \frac{qN_m}{\varepsilon_o \varepsilon_{ox}}\left[\overline{X}(V_G) - \overline{X}(-V_G)\right] \qquad (5.16)$$

where N_m is the mobile ionic charge density per unit area. Therefore, Eq. (5.13) becomes

$$\int\limits_{-V_G}^{V_G} [I_G - \alpha\, C_{ox}]dV_G = \alpha\, qN_m \left[\frac{\overline{X}(V_G)}{t_{ox}} - \frac{\overline{X}(-V_G)}{t_{ox}}\right] \qquad (5.17)$$

For most values of bias and time, the centroid in Eq. (5.15) is bias and time independent. Consequently, Eq. (5.15) shows that for linear voltage ramp, at which temperature is enough to make $C_{LF} = C_{ox}$, the area under an $I_G(V_G)$ curve in excess of $I_G = \alpha\, C_{ox}$ is proportional to mobile ion density per unit area. To use Eq. (5.15) to estimate N_m, one ordinarily assumes $\overline{X}(V_G) = t_{ox}$ (all positive ions drifted to the silicon surface) and $\overline{X}(-V_G) = 0$ (all ions drifted to the gate). These assumptions probably are valid to within 10 nm.

5.3.2 Earlier Investigation

Experimental determination of $I_G(V_G)$ curves have been carried out [9] which employ TVS technique. In such experiments, a heavy contaminated oxide with mobile ions is used. A typical $I_G(V_G)$ curves at 202°C is shown in Fig. 5.7. The sweep from positive to negative gate bias, when ions drift from the Si–SiO$_2$ interface to the metal-oxide interface is shown in the upper curve. However, the voltage sweep from negative to positive, when ions drift from the metal-oxide interface, is shown in the lower curve. The area of the peak above the C_{ox} baseline corresponds to 2.5×10^{12} ions/cm^2. The most striking feature of these curves is the strong asymmetry of the $I_G(V_G)$ curves in the two directions of the voltage sweep.

The ionic current response, when sweeping from the negative to positive gate bias, is distorted and broadened over the entire positive voltage range, and results in a very broad peak as shown in the lower curve of Fig. 5.7. This clearly indicates that ion transport is independent of a transport mechanism limited by the emission from ionic traps located at the metal-oxide interface. The areas under both peaks in Fig. 5.7 are identical so that all ions transported to metal-oxide interface can be brought to the silicon-oxide interface by changing the polarity of the voltage ramp.

Fig. 5.7 Gate current as function of gate voltage (Reproduced with permission, J. Electrochem. Soc. [9])

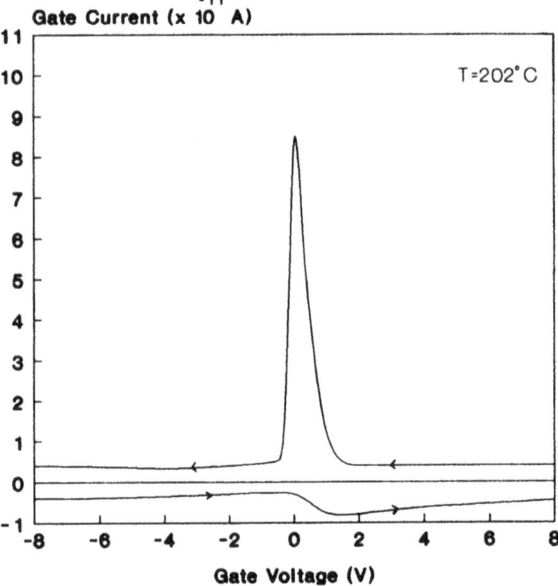

Fig. 5.8 Gate current as function of gate voltage for different temperature (Reproduced with permission, J. Electrochem. Soc. [9])

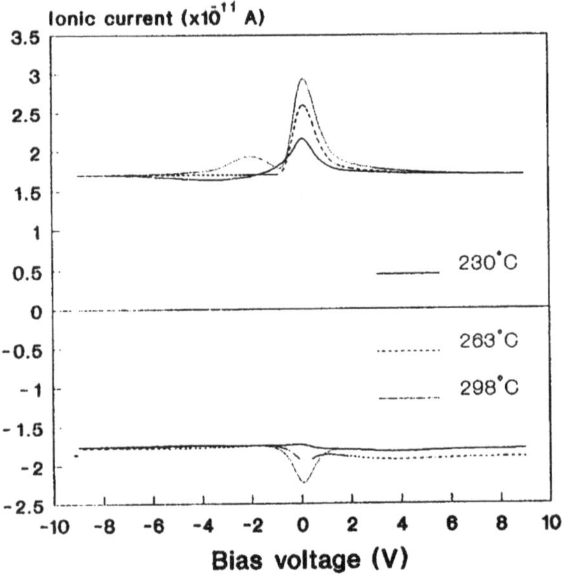

As expected from quasi-equilibrium arguments, the area under the peaks are independent of voltage sweep rate, as long as the rate is sufficiently slow to establish the boundary conditions (the mobile ions are concentrated as a thin sheet of charge at either interfaces). The temperature dependence of the ionic current response is illustrated in Fig. 5.8. The ionic current peak does not saturate but continues to increase with temperature. At low temperature, the sweep from negative to positive gate bias yields a broadened structure, resulting from the trapping emission mechanism at the metal-oxide interface. At higher temperatures, this peak sharpens and approaches were closely the ideal shape (or the shape of the peak for the opposite sweep) as expected if emission of ions from traps is more rapid at higher temperatures. Furthermore, an additional structure is observed above the C_{ox} baseline at elevated temperature ($T = 298°C$), as shown in the upper trace of Fig. 5.8. This structure typically takes the form of a second broadened peak displaced by approximately -2 V from the first peak. This second peak is due to the K^+ ions drift. It can be noted that the peaks due to potassium may not be visible at low temperature. Thus, the TVS technique might provide a simple means for identifying the density of sodium and potassium separately because the peak in the ionic displacement current occurs at a different gate bias [12].

5.4 TSIC Technique

The TSIC technique is usually applied to MOS capacitors that can be brought to an elevated temperature. As the temperature varies as a function of time, the current in the external circuit of the MOS capacitor is measured. In principle, the measured current is a superposition of the charging current of the MOS capacitor, caused by changes in temperature, and the ionic current (the displacement current) caused by moving ions. This ionic current has been studied in the past by a variety of methods [13]. Boudry and Stagg [14], and Hillen [15] have used TSIC measurements to study the kinetic behavior of mobile ions in MOS structure. The thermally stimulated currents were always measured after an appropriate BTS treatment (15 min at 350°C and a gate voltage +5 V). Before measurements at a positive gate voltage the stress voltage is initially kept negative (all ions are then driven to the oxide-metal interface), and for measurements at a negative gate voltage the stress voltage is initially kept positive. The starting temperature of the measurement must be low enough to ensure that reversing the electric field in the oxide will not cause any ionic current, meaning thereby that the ions are deeply trapped near the $Si–SiO_2$ interface (one usually uses a temperature of $-20°C$). In the experimental setup system (see Fig. 5.1) for measuring TSIC curves some tools such as HP4140, temperature controller ITC502 and cryostat have been used. Figure 5.9 shows the variations of the applied temperatures during TSIC experiment, in which ion motion from the $Si–SiO_2$ interface to the metal-SiO_2 interface is studied. During the current measurement, the temperature varies linearly with time. Typical results obtained on an MOS capacitor having Na^+ and K^+ ions in its oxide layer are also shown in this figure; two current peaks are shown in the temperature range 0–400°C, at a maximum applied field 1 MV/cm. These peaks are attributed to the motion of Na^+ and K^+ ions respectively.

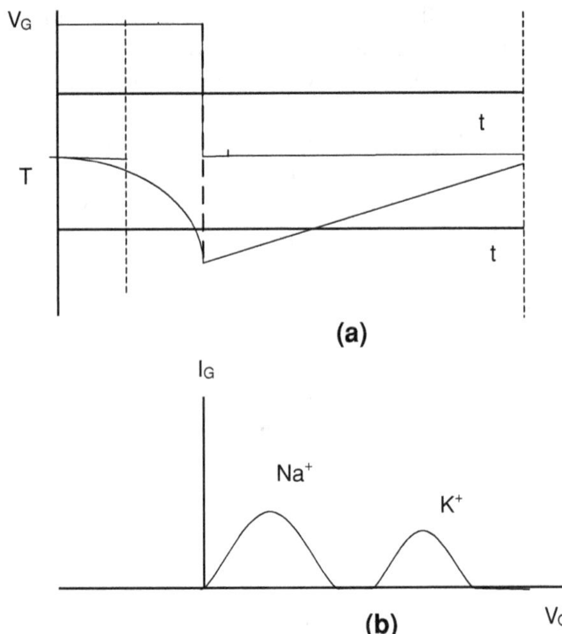

Fig. 5.9 a Variation of applied gate voltage during TSIC technique, **b** typical gate ionic current

5.4.1 Theory

The simplest model which explains the TSIC current states that, once the mobile ions are de-trapped after acquiring a sufficient energy they immediately move to the other side of the oxide edge where they are trapped again. Thus the thermally stimulated current is controlled (or limited) by the emission mechanisms of the ion traps located at either edge of the oxide layer.

Hickmott [13] explained several TSIC curves using a single level theory. The ionic current is expressed as:

$$I(t) = -qA\frac{dn(t)}{dt} \tag{5.18}$$

where

$$\frac{dn(t)}{dt} = -n(t) \cdot S \cdot \exp\left[-\frac{E_o}{kT(t)}\right], \tag{5.19}$$

and

$$n = n_o \qquad \text{at} \quad t = 0. \tag{5.20}$$

where n is the number of ions (per unit area) still trapped at time t, and E_o the activation energy required to excite an ion from a trap at the emitting interface to a point just outside the trap from which it is free to move through the oxide, and S a factor which may be roughly interpreted as the number of times per second an ion attempts to leave the trap. Since an ion is carried away from the interface by the applied electric field immediately after it has left the trap, it seems reasonable to ignore the possibility of its re-trapping in this situation. After taking n_o, E_o, and S as adjustable parameters, Hickmott [16] was able to obtain an adequate fit only to the leading slope and maximum of the negative peak and found E_o to be ranging from 0.7 to 1.2 eV and S from 10^7 to 10^{12}. The model describes the experimental curves well when a Gaussian spread in the activation value for the emission time constant is assumed (10^{-12} s). Further activation energy is found that is consistent with an emission-limited process [15, 20]. TSIC method complements BTS method for studying the metal-SiO_2 interface in MOS structure [16]. Stagg [21] also uses this method to determine the drift mobility of Na^+ and K^+ ions in SiO_2 films.

5.5 Charge-pumping Associated with BTS Technique

Charge-pumping technique is used for measuring the flat-band voltage shift that is due to BTS. This flat-band voltage shift can be used to determine the mobile ion concentration within the oxide layer in MOS device. The charge pumping current of the given MOS device is first measured before applying any stress. Then, the MOS structure is heated to a certain temperature up to 200°C and held there for a period up to 30 min, which is long enough to ensure that all the available ions drift completely across the oxide. At the same time, a positive gate bias is applied which is enough to produce an oxide electric field of a few MV/cm. After holding the MOS device at elevated temperature and high field for the required period, it is cooled back to room temperature so that no further redistribution of charge takes place during the second charge-pumping current measurement. The flat band voltage shift, which is obtained from a difference between the charge-pumping current curve before and after BTS, may be used to determine the mobile ions concentration drifted at the given temperature [17].

5.5.1 Theory

When a MOSFET transistor is pulsed into inversion, the surface region becomes deeply depleted and electrons will flow from the source and drain regions into the channel where some of them will be captured by the surface states. When the gate pulse is deriving the surface states back into accumulation, the mobile charges drift back to the source and drain under the influence of the reverse bias, but the charges trapped in the surface states will recombine with the majority carriers from the substrate. The charge Q_{it} that will recombine is given by [22]:

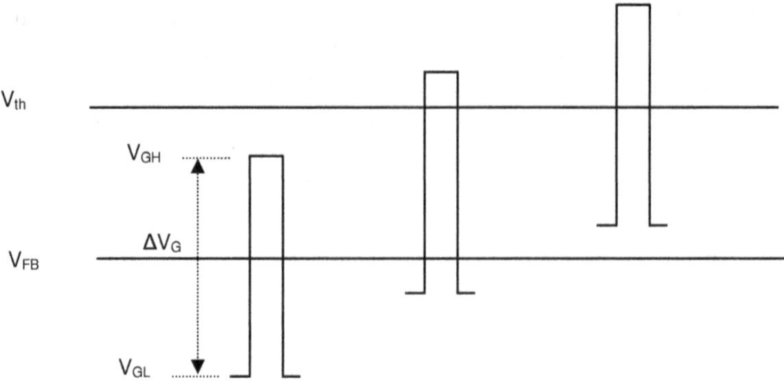

Fig. 5.10 Illustration of the base level method on a nMOSFET

$$Q_{it} = A_G q \int D_{it}(E)dE \tag{5.21}$$

where A_G is the area of the channel of the MOSFET (cm^2) and $D_{it}(E)$ is the interface trapped charge density at energy E.

When applying repetitive pulses to the gate with frequency f, varying the pulse base level from inversion to accumulation keeping the amplitude of the pulse constant as illustrated in Fig. 5.10, the charge Q_{it} *will give rise to a current in the substrate. This current can be expressed as:*

$$I_{cp} = f \cdot Q_{it} \tag{5.22}$$

By measuring this substrate current, an estimate of the mean capture cross-section, threshold voltage and the interface-state density over the energy range swept by the gate pulse can be obtained. Furthermore, the charge-pumping current I_{cp} undergoes an abrupt increase from zero to its maximum value after a certain position of the gate pulse base level. It has been found that the value of the base voltage corresponding to the half of maximum current $I_{cpm}/2$, in the I_{cp}–V_G curve is the flat-band voltage [23].

5.5.2 Separation of the Mobile Charge Effect

The previous work found that flat-band voltage shift is due to the change in the distribution of mobile ionic charge and trapped charge under the BTS. However, the oxide fixed charge distribution does not change under this treatment [3]. Whether this shift in the flat band is due to mobile ions or to traps, it can be written as [23].

$$\Delta V_{\text{FB}} = -\int_0^{t_{\text{ox}}} \frac{\rho(x)x\,dx}{\varepsilon_0\varepsilon_{ox}} \tag{5.23}$$

where $\rho(x)$ is the volume-density of charge within the oxide.

The effect of each charge layer depends on its distance from the oxide-silicon interface as given in Eq. (5.23). A layer has no effect if it is located at the metal-oxide interface and has a maximum effect if it is located at the oxide-silicon interface. If the charge contained in the oxide is only due to mobile ions Q_{m}, a change in the distribution, caused by an applied stress (temperature + electric field), gives rise to a new value of the flat-band voltage V_{FB}. This effect can be used to calculate the total amount of mobile ions or the density distribution of mobile ions. Since the charge contained in the oxide is not only due to mobile ions but also due to traps, a change in the density distribution due to an excitation (temperature, photons) causes also a shift in the V_{FB} value but in opposite polarity. This effect can be separated by using a change in the maximum charge pumping current ΔI_{cp}. After applying bias thermal stress to MOS device, the charges that are due to the mobile ions or to traps can be considered as a charges sheet located at the oxide-silicon interface. Under this specific situation, Eq. (5.23) becomes [21]

$$\Delta V_{\text{FB}} = \frac{(Q_{\text{m}} - \Delta Q_{\text{ot}})}{C_{\text{ox}}} \tag{5.24}$$

where the change in the oxide trapped charge ΔQ_{ot} can be calculated from Eq. (5.22) as follows

$$\Delta Q_{\text{ot}} = \frac{\Delta I_{\text{cp}}}{f} \tag{5.25}$$

The substitution of Eq. (5.25) into Eq. (5.24) gives the total amount of mobile charges

$$Q_{\text{m}} = \Delta V_{\text{FB}} C_{\text{ox}} + \frac{\Delta I_{\text{cp}}}{f} \tag{5.26}$$

It can be noticed that by measuring the change in the maximum charge pumping current and the shift of the flat band voltage, the amount of mobile ions can be obtained.

5.5.3 Experimental Results and Discussion

The experimental set up for measuring the charge-pumping current is illustrated in Fig. 5.1. A gate voltage signal of constant amplitude that can sweep the energy level from the deep accumulation to the deep inversion is applied to a MOSFET. The drain and source are tied together and connected to a dc reverse-biasing

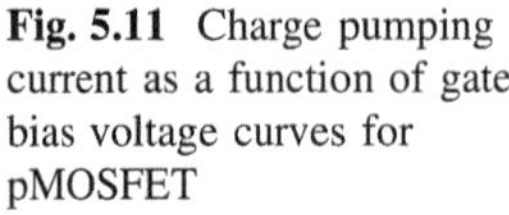

Fig. 5.11 Charge pumping current as a function of gate bias voltage curves for pMOSFET

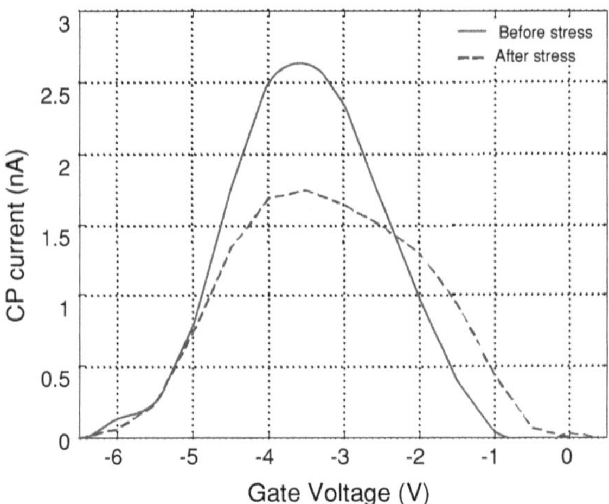

voltage [24]. A dc Pico-ammeter (HP4140B) is connected to a substrate to measure the resulting charge pumping current. Experimental measurements have been carried out on a number of commercial MOSFETs, such as n-MOSFET 3N171 and p-MOSFET 3N163, and encapsulated wafers in 64-pin PGA package with oxide thickness 40 and 12 nm respectively. The charge pumping current of the given MOS device is first measured before applying any stress. Then the MOS structure is heated to a certain temperature up to 200°C and held for a period up to 15 min, which is long enough to ensure that all the available ions at a given temperature drift completely across the oxide. At the same time a positive gate bias of 5 V is applied which is enough to produce an oxide field of 2 MV/cm. After holding the MOS device at elevated temperature and high electric field for the required period, it is cooled back to room temperature so that no further redistribution of charge takes place during the second charge-pumping current measurement. The flat band voltage shift between the charge-pumping current curve before and after bias-temperature stress is partially due to the mobile ion concentration drifted at a given temperature. The flat-band voltage shift may be used to determine the mobile ion concentration within the oxide layer in MOS device.

The flat band voltage shifts that are due to the BTS stress were measured by charge pumping as shown in Figures 5.11 and 5.12. These shifts are due to total effects of the change in the oxide charges that consists of a contribution of the mobile charges as well as a contribution of oxide trapped charges. It can be noted that there is a shift in the flat band voltage and a change in the maximum CP current magnitude. These phenomena can be used to compute the change in the oxide trapped charges density and the amount of mobile ions. This separation of mobile ion density from the oxide trapped charges density cannot be investigated in the case of HFCV technique. Through the use of the obtained curves of different MOSFET's and Eq. (5.26), the total mobile ions can be calculated as given in Table 5.1. Besides, this table shows values of the flat-band shift voltages for

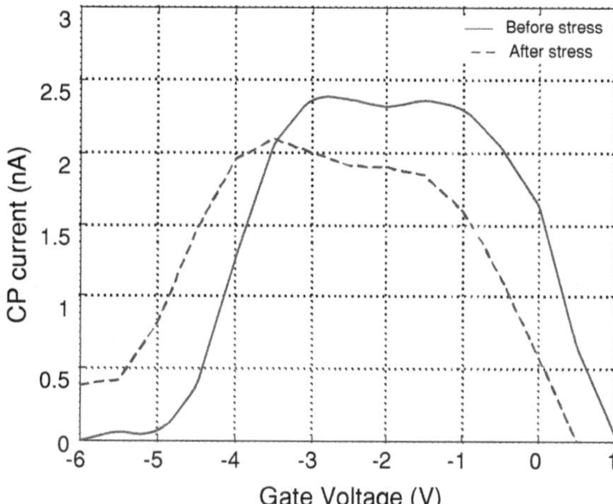

Fig. 5.12 Charge pumping current as a function of gate bias voltage curves for nMOSFET

Table 5.1 Values of the flat-band voltages shift and mobile ion quantity for different devices

Devices	ΔV_{FB} using CP	ΔV_{FB} using CV	ΔI_{cp} nA	$Q_{m\ 10}^{12}$ (cm^{-2})
pMOSFET	0.50 V	0.45 V	0.8	4.44
nMOSFET	0.62 V	0.50 V	0.4	4.75

different devices as obtained by the present method and their comparison with those obtained by the CV method.

The mobile ions density obtained by the measurement of the charge pumping current under BTS is quite reliable and are not affected by any error in the measurement of V_{FB} which may be caused by the trap level density at Si–SiO$_2$ interface. It is so because in charge pumping technique, any change in the trap level density can be measured separately and its effect can be subtracted from the measurement of V_{FB} shift. It may be noted that as the bias-temperature treatment causes small changes in interface trap level density, the maximum CP current before and after bias-temperature drift changes little bit. Besides, the shift in flat band voltage has a lower value when the same amount of ions is present in a thinner oxide.

References

1. Bentarzi, H., Zerguerras, A., Mitra, V.: The measurement techniques of the mobile ions in MOS structures. Algerian J. Techno. **11**, 19 (1995)
2. Nicollian, E.H., Brews, J.R.: MOS Physics and Technology. Wiley, New York (1982)
3. Hillen, M.W., Verwey, J.F.: Mobile ions in SiO$_2$ layers on Si. In: Barbottain, G., Vapaille, A. (eds.) Instabilities in Silicon Devices, pp. 404–439. Amsterdam (1986)
4. Sze, S.M.: Physics of Semiconductor Devices. Wiley, New York (1981)

5. Tangena, A.G., Middelhoek, J., DeRooij, N.F.: Influence of positive ions on the current-voltage characteristics of MOS structures. J. Appl. Phys. **49**, 2876–2879 (1978)
6. DiMaria, D.J.: Room-temperature conductivity and location of mobile sodium ions in the thermal silicon dioxide layer of a metal-silicon dioxide-silicon structure. J. Appl. Phys. **48**, 5149–5151 (1977)
7. Yamin, M.: Charge storage effects in silicon dioxide films. IEEE Trans. Elect. Dev. **ED-12**, 88–96 (1965)
8. Chou, N.J.: Application of triangular voltage sweep method to mobile charge studies in MOS structures. J. Electrochem. Soc. **118**, 601–609 (1971)
9. Kuhn, M., Silversmith, D.J.: Ionic contamination and transport of mobile ions in MOS structures. J. Electrochem. Soc.: Solid State Sci. **118**, 966–970 (1971)
10. Pepper, M., Eccleston, W.: Cation transport in SiO_2. Phys. Stat. Sol. (a) **12**, 199–207 (1972)
11. Przewlocki, H.M., Marciniak, W.: The triangular voltage sweep method as a tool in studies of mobile charge in MOS structures. Phys. Stat. Sol. (a) **29**, 265–274 (1975)
12. Derbenwick, G.F.: Mobile ions in SiO_2: potassium. J. Appl. Phys. **48**, 1127–1130 (1977)
13. Hickmott, T.W.: Thermally stimulated ionic conductivity of sodium in thermal. J. Appl. Phys. **46**, 2583–2598 (1975)
14. Boudry, M.R., Stagg, J.P.: The kinetic behavior of mobile ions in the Al-SiO_2-Si system. J. Appl. Phys. **50**, 942–950 (1979)
15. Hillen, M.W.: Dynamic behavior of mobile ions in SiO_2 layers. In: Partelides, S.T. (ed.) The Physics of SiO_2 and its Interface. Pergamon, New York (1978)
16. Hickmott, T.W.: Dipole layers at the metal-SiO_2 interface. J. Appl. Phys. **51**, 4269–4281 (1980)
17. Bentarzi, H., Zitouni, A., Kribes Y.: Oxide charges densities determination using charge-pumping technique with BTS in MOS structures. WSEAS Trans. Electron. 101–110 (2008)
18. Kriegler, R.J., Devenyi, T.F.: Direct measurement of Na^+ ion mobility in SiO_2 films. Thin Solid Films **36**, 435–439 (1976)
19. Hillen, M.W., Greeuw, G., Verweij, J.F.: On the mobility of potassium ions in SiO_2. J. Appl. Phys. **50**, 4834–4837 (1979)
20. Nauta, P.K., Hillen, M.W.: Investigation of mobile ions in mos structures using the TSIC method. J. Appl. Phys. **49**, 2862–2865 (1978)
21. Stagg, J.P.: Drift mobility of Na^+ and K^+ ions in SiO_2 films. Appl. Phys. Lett. **31**, 532–533 (1977)
22. Groeseneken, G., Maes, H.E., Beltran, N., De Keersmaecker, R.F.: A reliable approach to charge-pumping measurements in MOS transistors. IEEE Trans. Elect. Dev. **ED-31**, 42 (1984)
23. Grove, A.S.: Physics and Technology of Semiconductor Devices. Wiley, New York (1967)
24. Mitra, V., Benfdila, A., Bouderbala, R., Bentarzi, H., Amrouche, A.: Charge-extraction technique for studying the surface states in MOS devices. IEEE Trans. Elect. Dev. **ED-40**, 923 (1993)

Chapter 6
Theoretical Approaches of Mobile Ions Density Distribution Determination

6.1 Introduction

It is well known that the presence of the mobile ions in the oxide of MOS devices or thin films can greatly influence their electrical characteristics and so their stability. Although, a general feeling prevailed for some time that this problem can be avoided if the density of the mobile ions is reduced in the MOS structure by guttering or clean processing. Soon it had to be revived the fact that certain processes such as heat, X-rays, laser or plasma treatment of the MOS structures at any fabrication stage can either introduce fresh ions or reactivate those existing already in an inactive form in the oxide. Therefore, the study of the different aspects of the mobile ions in the oxide is still of considerable importance in MOS devices.

The fact that the mobile ions in the oxide of MOS structures can undergo a drift under the influence of thermal electric stress, led the scientists to form a general belief that these mobile ions can attain an equilibrium density distribution under a given temperature-bias condition. Accordingly, a number of theoretical attempts [1–4] have been developed to obtain this equilibrium density distribution of mobile ions from theoretical considerations. All these attempts have been made under certain simplifying assumptions and limitations. The simplifying assumptions, which are common to all the theoretical investigations, are:

1. The oxide layer of MOS structure is uniform, and its thickness is small in comparison to the electrode diameter, so that the problem may be considered one-dimensional.
2. There is no charge exchange between the oxide layer and electrodes (ideal blocking electrodes).
3. An uncompensated charge of positive mobile ions exists in the oxide layer. There is no charge generation and recombination in the oxide layer.
4. The changes in the applied electric field are taken such that the mobile ions are always able to attain their equilibrium distribution.

H. Bentarzi, *Transport in Metal-Oxide-Semiconductor Structures,*
Engineering Materials, DOI: 10.1007/978-3-642-16304-3_6,
© Springer-Verlag Berlin Heidelberg 2011

There are three prominent approaches [1, 3–5], which may be considered as the main landmarks in the theoretical development of the subject of mobile ions distribution in the insulating layers. These are reviewed in the next sections.

So far, a few methods are known to determine the mobile ion concentration from the experimental measurements in explicit form [6–9]. The present work also describes two empirical approaches for the determination of the equilibrium density of the mobile ions along the oxide thickness of a MOS structure.

Finally, a numerical approach is developed to simulate of mobile ions density distribution, by solving numerically a partial differential equation governing the ions kinetic.

6.2 Problem Formulation

The first theoretical attempt to tackle the problem of the mobile ion distribution in insulating layers was made by Chou [1]. In fact, he extended the flux equation which was used earlier by Snow et al. [6] to study the ion transport in insulating films. In the presence of concentration gradient and electric field ζ in the insulating film, the equation of ions-flux F can be written as

$$F(x,t) = -D\left(\frac{\partial N(x,t)}{\partial x}\right) + \mu_{\text{ion}} \zeta N(x,t) \tag{6.1}$$

where D is the diffusion coefficient, μ_{ion} is drift mobility of the mobile ions, and ζ is the electric field applied to MOS structure at a depth x as measured from the metal–oxide interface. A combination of Eq. 6.1 with equation of continuity, $(\partial N/\partial t) = -(\partial F/\partial x)$, gives the equation governing the time-dependent concentration of ions in the oxide,

$$\frac{\partial N(x,t)}{\partial t} = D\left(\frac{\partial^2 N(x,t)}{\partial x^2}\right) - \mu_{\text{ion}} \zeta\left(\frac{\partial N(x,t)}{\partial x}\right). \tag{6.2}$$

Under an applied voltage V_A, a static equilibrium of mobile ions is established when $(\partial N/\partial t) = 0$. Thus,

$$D\frac{\partial^2 N(x,t)}{\partial x^2} - \mu_{\text{ion}} \zeta\frac{\partial N(x,t)}{\partial x} = 0 \tag{6.3}$$

There are no mobile ions now across the boundaries because the electrodes block them. Therefore, for $x = 0$ and $x = t_{\text{ox}}$ the ions-flux $F(x, t)$ is null or,

$$D\left(\frac{\partial N(x,t)}{\partial x}\right) - \mu_{\text{ion}} \zeta(x)\left(N(x,t)\right) = 0 \tag{6.4}$$

where, t_{ox} is the oxide thickness.

6.3 Earlier Analytical Approaches

There are three prominent approaches [1, 3–5], which may be considered as the main landmarks in the theoretical development. Chou [1] developed the first earlier approach. Then, Tangena et al. [3] have developed the next theoretical attempt in this direction. Romanov et al. [4] developed another noteworthy analytical model. In these earlier attempts, they have derived the equilibrium distribution of mobile ions in the oxide of MOS structures through direct or indirect use of the ion transport equation that has been obtained in different investigations using different physical ideas. The ion transport equation so obtained is then coupled with the Poisson's equation in all the methods to get a differential equation in terms of either electric field or ionic current density.

6.3.1 Analytical Approach of Chou

The first theoretical attempt for determining the mobile ions distribution within the oxide layers was made by Chou [1]. In fact, he extended the flux Eq. 6.1 which was used earlier by Snow et al. [6] for studying the ion transport in insulating films. Integration of Eq. 6.4 between 0 and x and application of the boundary conditions mentioned in the previous section gives the following important differential equation,

$$D\left(\frac{dN(x,t)}{dx}\right) - \mu_{\text{ion}}\zeta(x)N(x,t) = 0 \qquad (6.5)$$

By virtue of the Einstein relationship, $(\mu/D) = (q/kT)$, Eq. 6.5 becomes

$$\frac{kT}{q}\left(\frac{dN(x,t)}{dx}\right) - \zeta(x)N(x,t) = 0. \qquad (6.6)$$

Standard substitution of $y = \ln(N(x))$ in Eq. 6.6 gives

$$\frac{kT}{q}\left(\frac{dy}{dx}\right) - \zeta(x) = 0, \qquad (6.7)$$

and

$$\frac{kT}{q}\left(\frac{d^2y}{dx^2}\right) = \frac{d\zeta(x)}{dx}. \qquad (6.8)$$

Combination of Eq. 6.8 with Poisson's equation,

$$\frac{d\zeta(x)}{dx} = \frac{q}{\varepsilon_0\,\varepsilon_{ox}}N(x), \qquad (6.9)$$

gives a new differential equation in the ion distribution $N(x)$ as

$$2\left(\frac{d^2y}{dx^2}\right) = \frac{1}{k_1}(N(x) - N_o),\tag{6.10}$$

where

$$k_1 = \frac{kT\varepsilon_o\varepsilon_{ox}}{2q^2}.\tag{6.11}$$

Setting $(dy/dx) = v$, Eq. 6.11 may be reduced to

$$k_1 d(v^2) = (e^y - 1)dy.\tag{6.12}$$

Using the frequently used relation,

$$2\left(\frac{d^2y}{dx_2}\right) = \frac{d}{dy}\left(\frac{dy}{dx}\right)^2.\tag{6.13}$$

Equation 6.12 yields

$$\int_{v_o}^{v} k_1 d(v^2) = \int_{y_o}^{y} (e^y - 1)\, dy,\tag{6.14}$$

or

$$k_1 v^2 = k_1 v_0^2 + e^y - e^{y_o} - (y - y_o),\tag{6.15}$$

or

$$\sqrt{k_1}\frac{dy}{dx} = \sqrt{k_1 v_0^2 + e^y - e^{y_o} - (y - y_o)}.\tag{6.16}$$

Integration of Eq. 6.16 results in

$$\int_{y_o}^{y} \frac{1}{\sqrt{k_1 v_0^2 + e^y - e^{y_o} - (y - y_o)}}dy = \frac{x}{\sqrt{k_1}}.\tag{6.17}$$

Equation 6.17 can not be integrated in a closed form. It must be numerically integrated by a trial-and-error scheme. However, even numerically Chou could not solve the differential equation to get any concrete mobile ions distribution curve. Therefore, his investigation can not give any idea about the final solution. Therefore, his analysis remains only suggestive rather than conclusive of giving concrete mobile ion density profiles. Moreover, Chou assumed the presence of immobile negative ions in his treatment. Przewlocki and Marciniak [2] elaborate Chou's method by considering only the presence of positive mobile ions but they did not add anything more concrete to the solution except for a little different mathematical procedure and analysis to solve the differential equation of Chou.

6.3.2 *Analytical Approach of Tangena et al.*

Tangena et al. [3] have developed the next theoretical attempt in this direction that needs mentioning. Although they have used a different physical idea to obtain the basic equation but the resulting equation is again the same as flux equation used by Chou. Tangena et al. argued that under equilibrium condition of density distribution, the electrochemical energy $\overline{E_{\text{ion}}}$ of an ion everywhere in the oxide should have the same value so that

$$\overline{E_{\text{ion}}} = E_{\text{o}} + q(V(x)) + kT \ln\left(\frac{N(x)}{N_{\text{o}}}\right) \tag{6.18}$$

where E_{o} denotes the energy level of the positive ions in the oxide, q the electron charge, $V(x)$ the electrical potential at x, k the Boltzmann constant, T the absolute temperature and N_{o} the total density of the energy states which is available for the ions in the oxide. In fact, the distribution of energy E corresponding to third term on the right hand side of Eq. 6.18 is supposed to be governed by Boltzmann law so that,

$$N(x) = N_{\text{o}} \exp\left[\frac{-E(x)}{kT}\right] \tag{6.19}$$

Equation 6.19 provides the value of the third term on the right hand side of Eq. 6.18. Differentiation of Eq. 6.18 gives

$$q\frac{dV(x)}{dx} + \frac{kT}{N(x)}\frac{dN(x)}{dx} = 0 \tag{6.20}$$

Substituting the electric field $\zeta(x)$ for the negative gradient of the potential, Eq. 6.20 becomes

$$\frac{kT}{q}\frac{dN(x)}{dx} - \zeta(x)N(x) = 0 \tag{6.21}$$

Equation 6.21 is combined with the Poisson's equation as given in Eq. 6.9. However Unlike Chou [1], Przewlocki and Marciniak [2], Tangena et al. eliminated $N(x)$ in the oxide and got

$$\frac{kT}{q}\frac{d^2\zeta(x)}{dx^2} - \zeta(x)\frac{d\zeta(x)}{dx} = 0. \tag{6.22}$$

Integration of Eq. 6.22 gives

$$\frac{2kT}{q}\frac{d\zeta(x)}{dx} - \zeta^2(x) = C_1, \tag{6.23}$$

where C_1 is an integration constant. Equation 6.23 can be converted into the following form:

$$\frac{2kT}{q} \int_{\zeta} \frac{\mathrm{d}\zeta(x)}{\zeta^2(x) + C_1} = \int_0^x \mathrm{d}x = x \tag{6.24}$$

Left hand side of Eq. 6.24 is a standard integral with three different solutions depending upon the value of the integration constant C_1. It is argued out that only the case $C_1 > 0$ is important whereas the other cases when $C_1 \leq 0$ that correspond to the situation when the ions are located at either of the interfaces. Tangena et al. [3] have solved therefore, Eq. 6.24 only under the case $C_1 > 0$ and obtained a solution

$$\frac{2kT}{q} \left[\frac{1}{\sqrt{C_1}} \arctan\left(\frac{\zeta(x)}{\sqrt{C_1}} \right) \right] + C_2 = x, \tag{6.25}$$

where C_2 is another integration constant. Equation 6.25 can be rewritten as

$$\zeta(x) = \frac{2akT}{q} \tan(ax + b), \tag{6.26}$$

where, $a = \frac{q\sqrt{C_1}}{2kT}$, and $b = \frac{q\sqrt{C_1}}{2kT} C_2$.

Combining Eq. 6.26 with Poisson's Eq. 6.9, the density distribution of mobile ions can be expressed by

$$N(x) = \frac{2kT\varepsilon_0\varepsilon_{ox}}{q^2} \frac{a^2}{\cos^2(ax + b)}. \tag{6.27}$$

The values of the two constants a and b have been obtained under the boundary conditions:

$$\int_0^{t_{ox}} \zeta(x)\mathrm{d}x = -V_A, \tag{6.28}$$

and

$$\varepsilon_0\varepsilon_{ox} \int_{\zeta(0)}^{\zeta(t_{ox})} \mathrm{d}\zeta = Q_{tot}. \tag{6.29}$$

Graphic solution of Eq. 6.27 is obtained for certain assumed value of device parameters which is shown in Fig. 6.1. However, the results as depicted by Fig. 6.1 suffer with an ambiguity that by increasing the gate voltage to more positive value, the curve is shifted towards the metal–oxide interface instead of towards the Si–SiO$_2$ interface. This may be attributed partly to the boundary

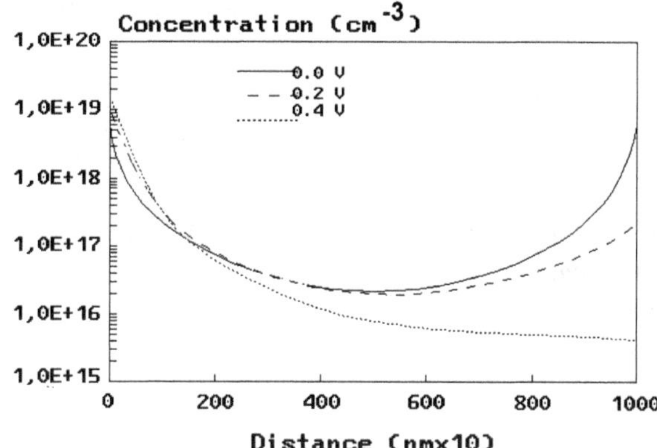

Fig. 6.1 The ion distribution $N(x)$ in the oxide at different applied voltages across the oxide ($T = 250°C$) (Reproduced with permission AIP [3])

conditions using assumed value of oxide voltage and oxide charge. In fact, the effective voltage in the oxide may be quite different from the applied voltage. Moreover, the method does not provide any explanation why only a certain part of the energy of the mobile ion is governed by Boltzmann law. Further, the method does not throw any light on the separate effect of different forces on the mobile ions.

6.3.3 Analytical Approach of Romanov et al.

Another noteworthy theoretical model is developed by Romanov et al. [4]. They have derived the equilibrium distribution of mobile ions in the oxide of MOS structures by commencing their theoretical treatment with the following current density equation,

$$j = -qD\frac{\mathrm{d}N(x)}{\mathrm{d}x} + q\mu_{\text{ion}}\zeta(x)N(x). \tag{6.30}$$

They considered in an oxide layer with boundaries impermeable to ions, the equilibrium state may take place only for $j = 0$. They combined Eq. 6.30 with Poisson's Eq. 6.9 and obtained the following relation,

$$\frac{\mathrm{d}^2\zeta(x)}{\mathrm{d}x^2} = \frac{q}{kT}\frac{\zeta(x)\,\mathrm{d}\zeta(x)}{\mathrm{d}x}. \tag{6.31}$$

In fact they solved Eq. 6.31 to get, at first, the following equation for the electric field distribution:

$$\zeta(x) = \sqrt{\frac{2kT}{q}C_1}\,\tan\left[\sqrt{\frac{C_1 q}{2kT}}(x + C_2)\right], \tag{6.32}$$

where C_1 and C_2 are the integration constants. When Eq. 6.32 is combined with Poisson's Eq. 6.9, it gives

$$N(x) = \frac{\varepsilon_0 \varepsilon_{ox} C_1}{q} \cos^{-2}\left[\sqrt{\frac{C_1 q}{2kT}}(x + C_2)\right].\tag{6.33}$$

The boundary conditions with $V(0) = V_A$, and $V(t_{ox}) = 0$ are used for determining the integration constants C_1 and C_2 as well as the total oxide charge Q_{tot} that is given by

$$Q_{tot} = q \int_0^{t_{ox}} N(x)dx.\tag{6.34}$$

For the case $C_1 > 0$ the integration constants C_1 and C_2 may be obtained using the following equations

$$\frac{Q_{tot}}{\varepsilon_0 \varepsilon_{ox}} = \sqrt{\frac{2kTC_1}{q}} \tan\left[\sqrt{\frac{C_1 q}{2kT}}(t_{ox} + C_2)\right] - \tan\left[C_2\sqrt{\frac{qC_1}{2kT}}\right],\tag{6.35}$$

$$V_A = \frac{2kT}{q}\ln\frac{\cos\left(C_3\sqrt{\frac{C_1 q}{2kT}}\right)}{\cos\left(\sqrt{\frac{C_1 q}{2kT}}(t_{ox} + C_2)\right)}.\tag{6.36}$$

Similar sets of equations giving the value of C_1 and C_2 have been obtained for the other cases corresponding to the conditions $C_1 \leq 0$. The value of the integration constants obtained in this way are used in Eq. 6.32 to obtain the density distribution profiles of the mobile ions for certain assumed value of V_A which are shown in Fig. 6.2. Although the obtained results by this method show more consistency with respect to the variation of the gate voltage as compared to previous models but still it lacks in providing precise quantitative results. For example the density profiles as shown in Fig. 6.2 correspond to certain assumed value of the oxide potential V_A. Therefore, this method also like others is not capable of giving precise value of the density of the mobile ions for a given device. Further, it also does not throw light on the different internal and external effects on the mobile ions separately.

All the above theoretical approaches are based on the direct or indirect use of the ion transport equation, which has been obtained in different investigations using different physical ideas. The ion transport equation so obtained is then coupled with the Poisson's equation in all the methods to get a differential equation in terms of either electric field or ionic current density. These methods, however, differ in their subsequent analysis of solving the differential equation to obtain the mobile ion density profiles.

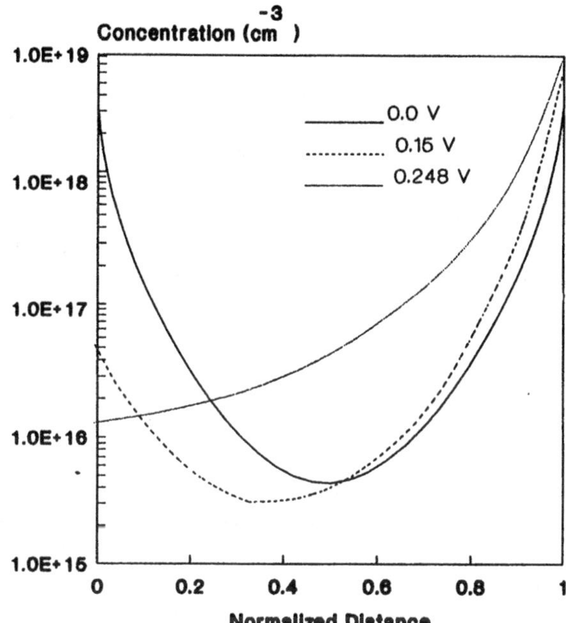

Fig. 6.2 The ion distribution $N(x)$ in the oxide at different applied voltages across the oxide (Reproduced with permission AIP [4])

6.4 Empirical Model

So far, a few methods are known to determine the mobile ion concentration from the experimental measurements in explicit form [6–10]. The present investigation describes two approaches for the determination of the equilibrium density distribution of the mobile ions along the oxide thickness of a MOS structure. In the first approach, an attempt is made with objective to determine mobile ion distribution simply from the knowledge of experimentally obtained values of the flat band voltage of a given MOS device under different conditions first before contamination, second after contamination and finally after drift of the ions under BTS stress [11]. In the second approach, a method for the determination of the equilibrium density distribution of the mobile ions has been described using experimentally measured values of its flat band voltage under different conditions namely before contamination/activation, after contamination/activation and then after ion drift due to thermal electric stress [12]. This is achieved by deriving an expression for the total ion concentration in terms of the flat band voltages under different cases of assumed distribution such as rectangular, exponential and Gaussian. The computed values of the flat band voltage shifts and the total ion concentration under different assumed distributions are then compared with experimental values. This is further supported by an additional computation using the analytical model [13] (see Chap. 7) developed by the authors.

6.4.1 General Formulation

Now, it is generally accepted that the flat-band voltage undergoes a shift whenever there is any change in the concentration or distribution of mobile ions within the oxide layer. A device that is fabricated under controlled conditions is supposed to have insignificant mobile ion concentration and such a device will be referred as controlled device. If such a controlled device is conditionally contaminated further with mobile ions, as mentioned in the previous section, the flat band voltage V_{FB} of the controlled device will change by an amount ΔV_{FB} due to the additional mobile ions given by [14]

$$\Delta V_{FB} = \int_0^{t_{ox}} \frac{\rho(x)x\,dx}{\varepsilon_0 \varepsilon_{ox}} \tag{6.37}$$

where $\rho(x)$ is the volume density of charge within the oxide, t_{ox} the oxide thickness, ε_0 the permittivity of the free space, ε_{ox} the relative permittivity of oxide and the distance x is measured from the metal–oxide interface.

The quantity ΔV_{FB} will depend upon the distribution of the mobile ions and hence on the volume ion density $\rho(x)$ within the oxide. The flat band voltage shift, ΔV_{FB}, can itself undergo a significant change if the ion distribution is changed somehow within the oxide. The value of ΔV_{FB} as determined immediately after intentional contamination/activation of mobile ions is due to the initial distribution of ions and will be referred as ΔV_{FB1},

$$\Delta V_{FB1} = \frac{1}{\varepsilon_{ox}\varepsilon_0} \int_0^{t_{ox}} x\rho(x)\,dx. \tag{6.38}$$

By applying an external electric field at elevated temperature [15], the mobile ions will undergo a drift towards the Si–SiO$_2$ interface and the resulting flat band voltage shift will be referred as ΔV_{FB2} and given as:

$$\Delta V_{FB2} = \frac{1}{\varepsilon_{ox}\varepsilon_0} \int_0^{t_{ox}} x\rho(-(x-t_{ox}))\,dx. \tag{6.39}$$

In all previous studies, the quantity ΔV_{FB1} has been ignored by assuming all the contaminated mobile ions to be concentrated in a thin sheet at the metal–oxide interface because it will not affect the value of flat band voltage before contamination [7, 15]. These ions have been again assumed to form a thin sheet of charge at the Si–SiO$_2$ interface after the drift which gives rise to different value of flat band voltage and hence a different value of ΔV_{FB2}. However, the present study is based on the contention that the mobile ions should obey certain equilibrium distribution before as well as after the drift rather than being concentrated in a thin sheet near either of the interfaces, because such an equilibrium distribution is bound to occur due to several types of internal forces acting upon them. This equilibrium distribution is likely to have the same form before and after the drift. As nothing changes from one case to another except the direction of concentration gradient. In fact, the effect of the thermal–electric stress may be considered as if the mobile ions are introduced from the Si–SiO$_2$ face instead of the other.

Any way, the total density of the ions Q_{tot} will remain constant within the oxide before and after the drift and which is given by

$$Q_{\text{tot}} = \int_0^{t_{\text{ox}}} \rho(x)\,dx \tag{6.40}$$

6.4.2 First Empirical Model

Regarding the distribution of mobile ions as discussed above, we consider in the present study the following general form:

$$\rho(x) = \rho(0) \exp\left(\frac{-X^n}{A}\right), \quad \text{for} \quad 0 < x < t_{\text{ox}} \tag{6.41}$$

where A, B and n are constants, and x is measured from the metal–oxide interface. Equation 6.41 as applied to the mobile ion distribution before and after drift becomes

$$\rho(x) = qN_{\text{o}} \exp\left(\frac{-X^n}{A}\right), \tag{6.42a}$$

and,

$$\rho(x) = qN_{\text{o}} \exp\left[\frac{(-1)^{n+1}(x - t_{\text{ox}})^n}{A}\right]. \tag{6.42b}$$

Substitution of the expression of $\rho(x)$ from Eq. 6.42a in Eqs. 6.38 and 6.40, gives the following relations

$$\Delta V_{\text{FB1}} = \frac{qN_{\text{o}}}{\varepsilon_{ox}\varepsilon_{\text{o}}} \int_0^{t_{\text{ox}}} x \cdot \exp\left(\frac{-x^n}{A}\right) dx, \tag{6.43}$$

and

$$Q_{\text{tot}} = qN_{\text{o}} \int_0^{t_{\text{ox}}} \exp\left(\frac{-x^n}{A}\right) dx. \tag{6.44}$$

Substitution of Eq. 6.43 in Eq. 6.44 yields

$$\Delta V_{\text{FB1}} = \frac{Q_{\text{tot}} \int_0^{t_{\text{ox}}} x \cdot \exp\left(\frac{-x^n}{A}\right) dx}{\varepsilon_{ox}\varepsilon_o \ \int_0^{t_{\text{ox}}} \exp\left(\frac{-x^n}{A}\right) dx}. \tag{6.45}$$

In a similar way, the flat-band voltage shift ΔV_{FB2} can be obtained,

$$\Delta V_{\text{FB2}} = \frac{Q_{\text{tot}} \int_0^{t_{\text{ox}}} x \cdot \exp\left[\frac{(-1)^{n+1}(x-t_{\text{ox}})^n}{A}\right] dx}{\varepsilon_{ox}\varepsilon_o \ \int_0^{t_{\text{ox}}} \exp\left[\frac{(-1)^{n+1}(x-t_{\text{ox}})^n}{A}\right] dx}. \tag{6.46}$$

From the measured values ΔV_{FB1} and ΔV_{FB2} of flat-band voltage shift before and after drift and total mobile charge Q_{tot}, the distribution parameters n and A can be obtained.

6.4.3 Results and Discussions

The experimental measurements, needed for computation in the present study, consist of high frequency C–V characteristics of a MOS structure under three cases as mentioned earlier, beside the determination of the total mobile ionic charge density Q_{tot}. In the present study, the experimental data of the previous works [9, 16] is being used for better reliability and comparison of the results. For example Snow et al. [16] have obtained high frequency C–V characteristics of a MOS diode before contamination and then after its contamination using rinse of dilute NaCl solution. The final characteristics were obtained after ions-drift in the contaminated device. They determined the total ion density Q_{tot} from the direct charge measurement as well as the saturation value of the flat band voltage shift under thermal–electric stressing. However, Raychaudhuri et al. [9] have used neutron activation to increase the mobile ion concentration in the controlled device. The high frequency C–V curves of both the above mentioned measurements are shown in Fig. 6.3, in which curve before contamination or activation is marked as (a) then after contamination or activation as (b) and the final curve after drift as (c). It may be mentioned that ΔV_{FB1} and ΔV_{FB2} have been obtained by measuring the relative shifts between the curves (b) and (a) and between (c) and (a) at the flat band capacitance in both cases. Values of different quantities such as ΔV_{FB1}, ΔV_{FB2} and Q_{tot} as obtained above are shown in Table 6.1 along with the calculated distribution parameters A and n. Computation of the present study reveals, that the value of the index n comes out to be exactly the same and equal to 2 for both the measurements in spite of quite varying device-parameters such as oxide thickness t_{ox} and total ion concentration Q_{tot}. The value of $n = 2$ would mean a Gaussian distribution in conformity with the results of other studies [5, 13]. Further, the other distribution parameter A is also found to be nearly the same. The results of the present study are, therefore, conclusive that mobile ions within the oxide of a MOS device distribute

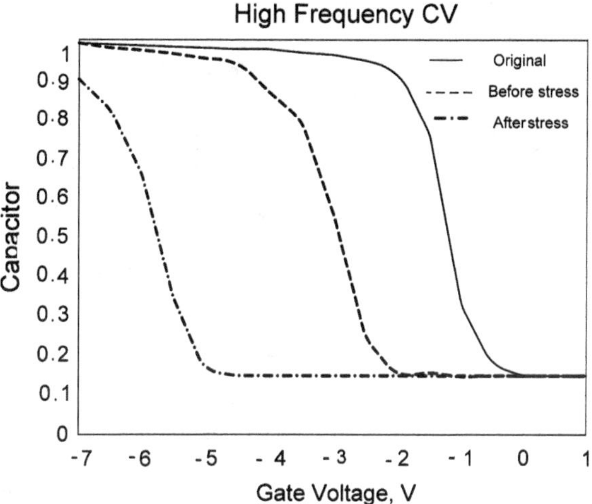

Fig. 6.3 Capacitance–voltage (CV) curves of MOS structure (Reproduced with permission IEEE [9])

Table 6.1 The calculated distribution parameters

Reference	Q_{tot} (10^{-12} cm^{-2})	ΔV_{FB1} (V)	ΔV_{FB2} (V)	t_{ox} (nm)	n	A (10^{-16} m^{-2})
[6]	2.9	1.4	28	200	2	57
[9]	1.8	1.6	4.2	35	1.8	126

themselves in a Gaussian manner both before as well as after the thermal–electric stress as shown in Fig. 6.4.

6.4.4 Second Empirical Model

In this model, the calculation of the theoretical flat band voltage shifts using Eqs. 6.38 and 6.39 is based on various charge profile considerations.

Uniform Rectangular Distribution

The uniform rectangular distribution of mobile charge is given by

$$\rho(x) = qN_0, \quad \text{for} \quad 0 < x < x_1, \tag{6.47a}$$

$$\rho(x) = 0, \quad \text{for} \quad x > x_1. \tag{6.47b}$$

Such charge distribution is illustrated in Fig. 6.5, which also explains the meaning of parameters N_o and X_1.

Fig. 6.4 Comparison of the empirical and theoretical model of concentration profile of mobile ions within the oxide of 35 nm thickness

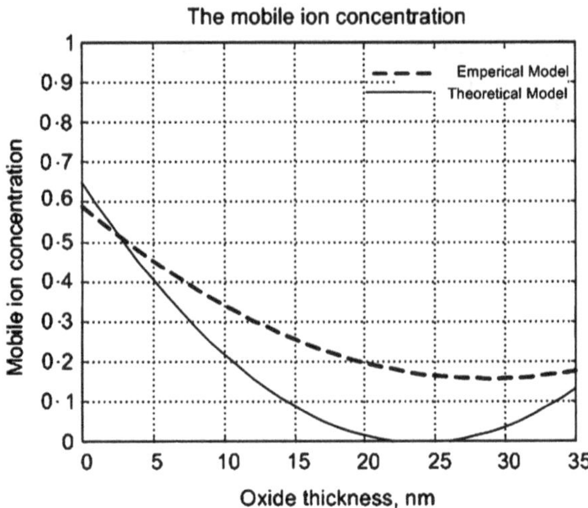

Fig. 6.5 Uniform rectangular distribution of mobile charge

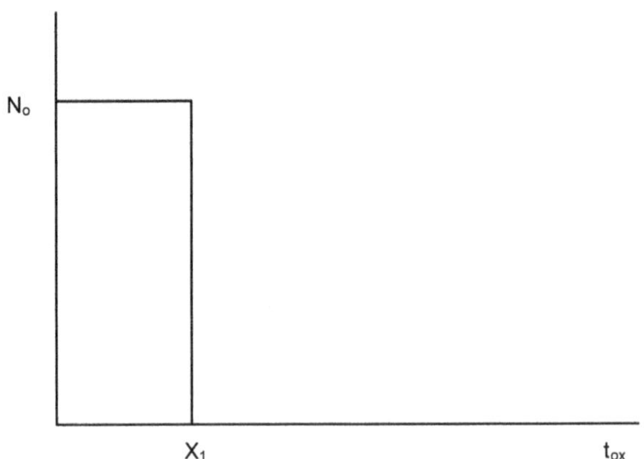

The combination of the three Eqs. 6.38, 6.40 and 6.47 gives

$$\Delta V_{\text{FB1}} = \frac{Q_{\text{tot}} x_1}{2\varepsilon_{ox}\varepsilon_0}. \tag{6.48}$$

Using the same procedure, the expression of second flat band voltage shift takes the following form

$$\Delta V_{\text{FB2}} = \frac{Q_{\text{tot}}(2t_{\text{ox}} - x_1)}{2\varepsilon_{ox}\varepsilon_0}. \tag{6.49}$$

Solving simultaneously Eqs. 6.48 and 6.49, the total ions and hence the charge distribution parameters N_0 and X_1 can be obtained

$$Q_{\text{tot}} = \varepsilon_0\varepsilon_{ox}\left(\frac{\Delta V_{\text{FB1}} + \Delta V_{\text{FB2}}}{t_{\text{ox}}}\right), \tag{6.50}$$

and

$$x_1 = \frac{2\Delta V_{FB1} t_{ox}}{\Delta V_{FB1} + \Delta V_{FB2}}. \tag{6.51}$$

The Exponential Distribution

The exponential charge distribution is given by

$$\rho(x) = qN_0 \cdot \exp\left(\frac{-x}{A}\right), \quad \text{for} \quad 0 < x < t_{ox}. \tag{6.52}$$

The exponential profile is shown in Fig. 6.6 that also gives the meaning of distribution parameters N_o and A.

In similar way as previous section, the expressions of the flat band voltage shifts can be obtained which come out to be

$$\Delta V_{FB1} = \frac{Q_{tot}}{\varepsilon_0 \varepsilon_{ox}}\left[A - \frac{t_{ox}}{\exp\left(\frac{t_{ox}}{A}\right) - 1}\right], \tag{6.53}$$

$$\Delta V_{FB2} = \frac{Q_{tot}}{\varepsilon_0 \varepsilon_{ox}}\left[-A + \frac{t_{ox}}{1 - \exp\left(\frac{t_{ox}}{A}\right)}\right]. \tag{6.54}$$

The two equations can be solved simultaneously for obtaining the distribution parameters N_o and A if the flat band shifts are known (e.g. from measurements).

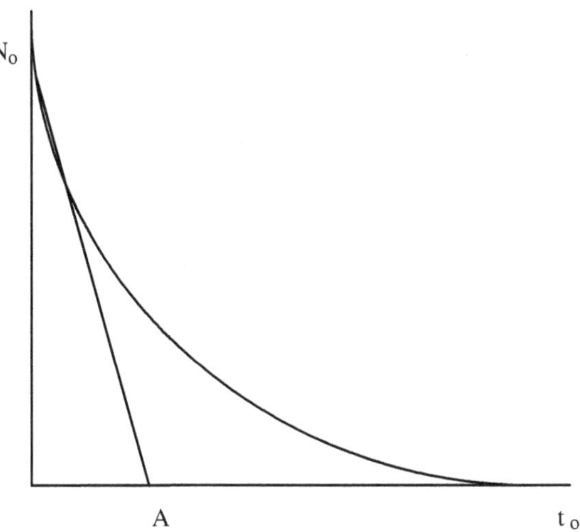

Fig. 6.6 Exponential charge distribution of mobile ions in the oxide

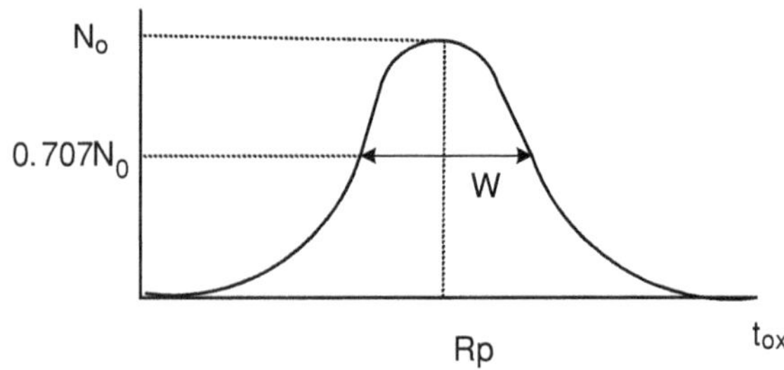

Fig. 6.7 Gaussian distribution of mobile ions in oxide of MOS structure

The Gaussian Distribution

The Gaussian distribution of the ionic charge is given by

$$\rho(x) = qN_0 \cdot \exp\left[-\left(\frac{x - R_p}{\sqrt{2}W}\right)^2\right], \quad \text{for} \quad 0 < x < t_{ox}, \tag{6.55}$$

and illustrated in Fig. 6.7, which also explains the meaning of the distribution parameters N_o and R_p and W. It is assumed here that $R_p = 0$, when mobile ions are at the metal–oxide interface, and $R_p = t_{ox}$, when mobile ions are drift to the silicon-oxide interface.

Under the above-mentioned conditions, the combination of Eqs. 6.55, 6.38 and 6.39 gives

$$\Delta V_{FB1} = \frac{Q_{tot} \int_0^{t_{ox}} x \cdot \exp\left(-\frac{x^2}{2W^2}\right) dx}{\varepsilon_0 \varepsilon_{ox} \int_0^{t_{ox}} \exp\left(-\frac{x^2}{2W^2}\right) dx}, \tag{6.56}$$

and,

$$\Delta V_{FB2} = \frac{Q_{tot} \int_0^{t_{ox}} x \cdot \exp\left(-\frac{(x-t_{ox})^2}{2W^2}\right) dx}{\varepsilon_0 \varepsilon_{ox} \int_0^{t_{ox}} \exp\left(-\frac{(x-t_{ox})^2}{2W^2}\right) dx}. \tag{6.57}$$

Thus, in order to find the total ions Q_{tot} and the distribution parameters N_o and W, Eqs. 6.56 and 6.57 should be solved numerically.

6.4.5 Results and Discussion

In the present study, the experimental data of certain earlier works [9, 16] has been used which can fulfill the required conditions. For example, Snow et al. [16] have obtained the high frequency C–V characteristics of a MOS diode before contamination, after its contamination using rinse of dilute NaCl solution and finally after the ion drift in the contaminated device. However, Raychaudhuri et al. [9]

Table 6.2 The measured values

Reference	t_{ox} (nm)	Q_{tot} 10^{-12} cm^{-2}	ΔV_{FB1} (V)	ΔV_{FB2} (V)
[6]	200	2.9	1.4	28
[9]	35	4.0	1.6	4.5

Table 6.3 The calculated distribution parameters

Reference	Rectangular		Gaussian		Exponential	
	X_1 (nm)	Q_{tot} (10^{-12} cm^{-2})	W (nm)	Q_{tot} (10^{-12} cm^{-2})	A (nm)	Q_{tot} (10^{-12} cm^{-2})
[6]	19	3.17	12.5	3.15	10	2
[9]	18.5	3.76	13	3.89	11	0.66

have used activation by ion beams to increase the mobile ion concentration in the controlled device. Values of different quantities such as ΔV_{FB1}, ΔV_{FB2} and Q_t as obtained above are shown in Table 6.2.

The measured flat band voltage shifts were used to determine the distribution parameters for different types of charge distribution. To prove the ability of this approach, the obtained distribution parameters for various possible distributions that are given in Table 6.3, are compared to experimental results [9, 16] as in Table 6.1. It is clear in Table 6.3 that the obtained distribution parameters for Gaussian distribution and uniform rectangular distribution approximates better the experimental results than the distribution parameters calculated for exponential distribution. However, the Gaussian appears to be quite realistic.

6.5 Numerical Approach

All the above theoretical approaches differ in their subsequent analysis of solving the differential equation to obtain the mobile ion density profiles. Nevertheless, none is capable of giving precise value of the density of the mobile ions for a given device. Further, the density distribution of mobile ions can be obtained at only equilibrium state. Besides, some parameters such as the mobility and diffusion coefficient have been considered constant even under BTS.

The differential Eq. 6.4 cannot be analytically integrated in a closed form. However, our study has suggested solving Eq. 6.4 numerically by Crank–Nicholson method [17]. Complication rises in the solution of this equation because the electric field ζ, diffusion coefficient D and the mobility μ_{ion} are not constant. When a voltage V_A is applied, the electric field becomes [18]

$$\zeta(t) = \frac{V_A}{t_{ox}} + \frac{qN_o}{\varepsilon_0\varepsilon_{ox}}\left(1 - \frac{\bar{X}(t)}{t_{ox}}\right), \tag{6.58}$$

where N_o is the total number of ions and $N(t)$ is the density of ions as function of time at the centroid $\bar{X}$.

The diffusion coefficient D and mobility μ_{ion} of mobile ions in thermally grown Silicon-Oxide can be obtained for a given temperature from the following expressions [14]

$$D(T) = D_0 \cdot \exp\left[\frac{E_A}{E_B}\left(\frac{1}{T_0} - \frac{1}{T}\right)\right], \tag{6.59}$$

and,

$$\mu_{ion}(T) = \mu_0 \exp\left[-\frac{E_A}{kT}\right]. \tag{6.60}$$

where E_A is the activation energy. According to papers [7, 14] the value of D_0 at $T = 140°C$ is 3.41×10^{-10} cm^{-2}/s, $E_A = 0.66$ eV, and $\mu_0 = 1.05$ cm^2/V·s for T between $40°$ and $180°C$.

The initial condition and boundary conditions have to be assumed to fulfil both mathematical and physical considerations. The initial condition corresponding to $t = 0$, which is taken when all sodium ions are at the metal–oxide interface, is given as

$$N(x, 0) = N_o \cdot \exp\left[-\left(\frac{x}{\sqrt{2W}}\right)^2\right], \tag{6.61}$$

where $W = \frac{t_{ox}}{5}$.

This condition can be easily obtained experimentally by applying an appropriate external electric field to the MOS structure.

The parameter that can be greatly affected by the redistribution of mobile ions is the flat band voltage. The resulted shift in the flat band voltage that is due to change in the ion density distribution can be given as

$$\Delta V_{FB} = \frac{q}{\varepsilon_0 \varepsilon_{ox}} \int_0^{t_{ox}} xN(x, t)dx. \tag{6.62}$$

It is now generally accepted that the flat band voltage undergoes a shift whenever there is any change in the distribution of the mobile ion density within the oxide layer. This shift can be used as verification means.

6.5.1 Numerical Solution

The numerical solution of Eq. 6.2 with help of Eqs. 6.4 and 6.61 is done by using the finite difference method defined on different grid points in the oxide layer as shown in Fig. 6.8. The Crank–Nicholson representation of Eq. 6.2 is:

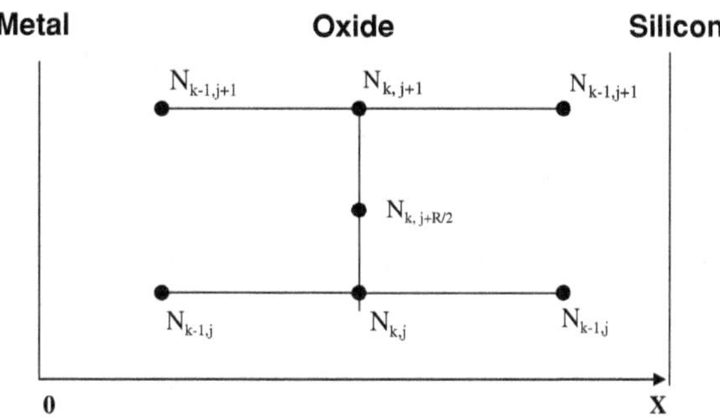

Fig. 6.8 Grid points representation of a MOS structure

$$N_{j+1,k-1} - \frac{2H^2}{D}\left(\frac{D}{H^2} + \frac{1}{R}\right)N_{j+1,k} + N_{j+1,k+1}$$

$$= -N_{j,k-1} + \frac{2H^2}{D}\left(\frac{D}{H^2} - \frac{\mu_{\mathrm{ion}}\zeta}{H} - \frac{1}{R}\right)N_{j,k}$$

$$+ \frac{2H^2}{D}\left(\frac{\mu_{\mathrm{ion}}\zeta}{H} - \frac{D}{2H^2}\right)N_{j,k+1} \quad \text{for} \quad j = 1, 2, \ldots n; \quad \text{and} \quad k > 0. \tag{6.63}$$

where $H = \Delta x$, $R = \Delta t$ and n is the number of divisions in the x direction, the counter k is used for the x-direction (distance), whereas the counter j is for the t-direction (time). When it is written for the entire difference grid, implicit formulas generate sets of linear algebraic equations whose solutions can be obtained by using MATLAB. Then, the Eq. 6.63 can be rewritten in the form of a tri-diagonal linear system $AX = B$ as follows:

$$\begin{bmatrix} (\lambda - \alpha) & 1 & 0 & . & 0 \\ 1 & -\alpha & 1 & 0 & . \\ 0 & 1 & -\alpha & 1 & 0 \\ . & . & . & . & . \\ 0 & . & 0 & 1 & (\frac{1}{\lambda} - \alpha) \end{bmatrix} \begin{bmatrix} N_{j+1,1} \\ N_{j+1,2} \\ . \\ . \\ N_{j+1,n} \end{bmatrix} = \begin{bmatrix} \phi_1 \\ \phi_2 \\ . \\ . \\ \phi_n \end{bmatrix} \tag{6.64}$$

where,

$$\alpha = \frac{2H^2}{D}\left(\frac{D}{H^2} + \frac{1}{R}\right), \quad \beta = \frac{2H^2}{D}\left(\frac{D}{H^2} - \frac{\mu_{\mathrm{ion}}\zeta}{H} - \frac{1}{R}\right), \quad \gamma = \frac{2H^2}{D}\left(\frac{\mu_{\mathrm{ion}}\zeta}{H} - \frac{D}{H^2}\right),$$

$$\lambda = \left(\frac{D}{\mu_{\mathrm{ion}}\zeta H + D}\right), \quad \text{and} \quad \phi_k = -N_{j,k-1} + \beta N_{j,k} + \gamma N_{j,k+1}.$$

The initial condition $N(x, t)$ can be approximated by:

$$N_{1,k} = N_0 \cdot \exp\left[-\left(\frac{kH}{\sqrt{2}W}\right)^2\right], \quad \text{for} \quad kH \leq t_{\mathrm{ox}}. \tag{6.65}$$

When an approximation of the boundary condition is used at the left, Eq. 6.4 becomes:

$$N_{j,0} = \frac{D}{\mu_{\text{ion}}\zeta H + D} N_{j,1}.$$ (6.66)

This permits to modify Eq. 6.63 at this boundary to

$$(\lambda + \alpha)N_{j+1,1} + N_{j+1,2} = (-\lambda + \beta)\; N_{j,1} + \gamma\, N_{j,2}.$$ (6.67)

Similarly, the same condition at the right boundary transforms Eq. 6.63 to

$$\left(\frac{1}{\lambda} - \alpha\right)N_{j+1,n} + N_{j+1,n-1} = \left(\beta + \frac{\gamma}{\lambda}\right)\; N_{j,n} - N_{j,n-1}.$$ (6.68)

The Crank–Nicholson method has been chosen to solve Eq. 6.2 for its higher stability properties, higher accuracy, and it is more rapid to reach the equilibrium distribution of mobile ions in the oxide layer.

6.5.2 Simulation Results and Discussion

A computer program has been developed and variety of simulated mobile ionic charge profiles under different conditions have been obtained for various parameters such as temperature, external electric field. The obtained results support the experimental data and they agree already published results [8, 9, 12, 15–17].

The developed program has been used to obtain the concentration-profile (N–x curve) of the mobile ions along the oxide-thickness. Typical plots for the mobile ion density-distribution, as obtained with the help of Eq. 6.42a, b, are shown in Fig. 6.9 at temperature $T = 27°C$, total density of ions $N_{\text{o}} = 3 \times 10^{11}$ cm^{-2}, and oxide-thickness 12 nm, for bias voltage V_A ranging from 0 V to 2 V for the biasing time $t = 0.0162$ h. When the MOS structure acquires enough positive

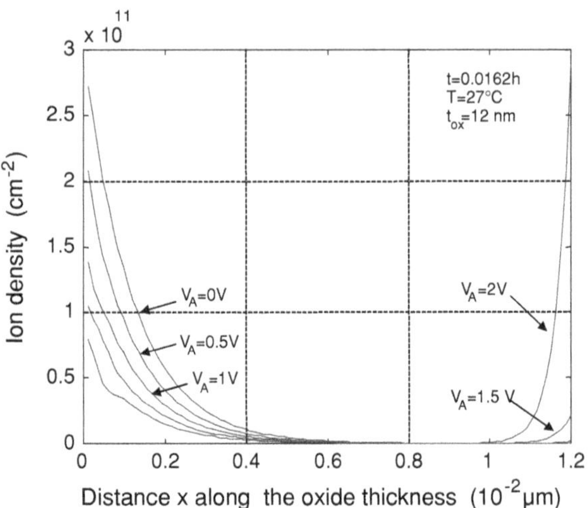

Fig. 6.9 The effect of bias voltage on mobile ions concentration within oxide

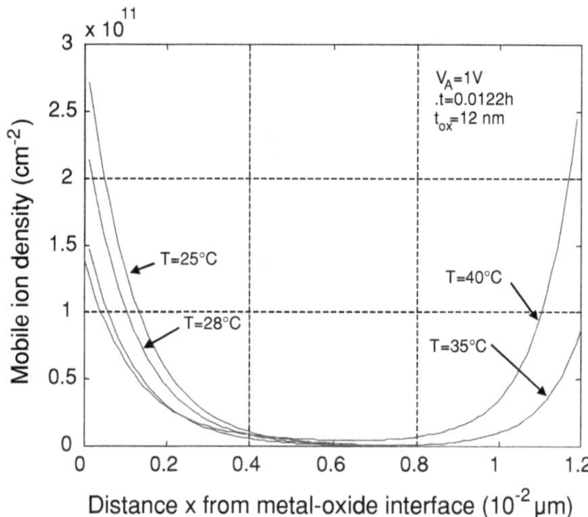

Fig. 6.10 The effect of temperature on mobile ions concentration

value, the mobile ions will move from the metal–oxide interface to oxide-silicon interface faster. Therefore, the density-distribution of the mobile ions that is initially Gausian falls rapidly. However, as the applied voltage V_A becomes more and more positive, the concentration of the ions starts building up at the SiO_2–Si interface and the distribution-curve takes a U-shape. Finally when V_A attains enough positive value, the distribution-curve shows an exponential increase of the ions towards the Si–SiO_2 interface. Figure 6.10 shows different distributions of ion density that are obtained by simulation for a certain biasing time $t = 0.0122$ h and for different temperatures. It can be noted that when the temperature increases the ions are accelerated and reach faster their equilibrium state. Since, the internal electric field that depends on the ion density distribution depends on the time. This internal electric field which has been expressed in the second term of the right side of Eq. 6.58 can be plotted as function of the time (see Fig. 6.11). It can be noticed that the effects of ions being pushed to the oxide layer boundaries and the possibility of the existence of intensive electric fields in high intensity regions have been theoretically substantiated.

6.5.3 Experimental and Simulation Results

Silicon p-type structures of $\langle 100 \rangle$ orientation with resistivity about 10 ohm-cm are used here. After the standard cleaning process, thin oxide layers of 120 Å were thermally grown at 1100°C in dry oxygen ambient diluted with nitrogen gas. Post oxidation annealing was implemented at 450°C in pure nitrogen ambient for 30 min. Then, an aluminium layer of 1000 Å thickness was deposited on the oxide layer using the EDWARDS 306A evaporator. The Bias Thermal Stress (BTS)

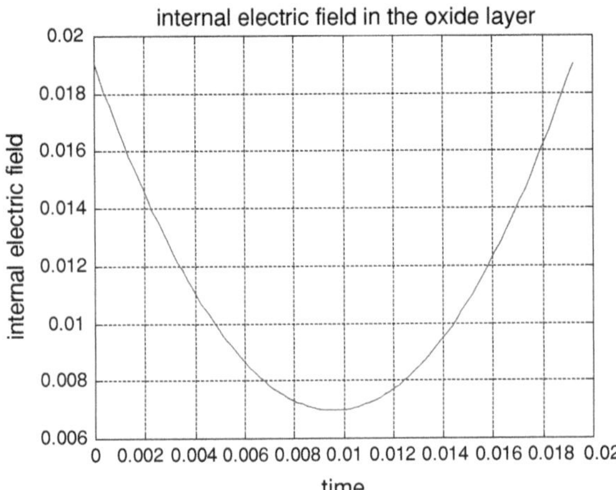

Fig. 6.11 The variation of the internal electric field (MV/Cm) as a function of time (h)

technique was performed through the use of HP 4275A LCR meter. It has been found experimentally that for an applied voltage of 1.5 V to the same MOS structure mentioned above, the flat band voltage shift ΔV_{FB} is 170 mV. In the other hand, for the same parameters and using our computer program, the flat band voltage shift ΔV_{FB} that has been found is 167.41 mV. Therefore, the computed flat band voltage shift is in good agreement with the experimental one.

6.6 Conclusion

Most of the previous approaches are based on the direct or indirect use of the ion transport equation that has been obtained in different investigations using different physical ideas. The obtained differential equation is either not solved completely as in the approach [1, 2] or it is solved under certain assumed initial and boundary conditions using assumed value of parameters such as electric potential, electric field or charge in the oxide [3–5]. Thus, none of the studies is capable of giving precise distribution for a given device under a given temperature-bias condition.

So far, a few methods are known to determine the mobile ion concentration from the experimental measurements in explicit form [6–9]. The present investigation describes two approaches for the determination of the equilibrium density distribution of the mobile ions along the oxide thickness of a MOS structure. In the first approach, the attempt is made with objective to determine mobile ion distribution simply from the knowledge of experimentally obtained values of the flat band voltage of a given MOS device under different conditions first before contamination, second after contamination and finally after drift of the ions under BTS stress [11]. Experimental data of previous investigations have been used and all of them yield the Gaussian distribution. In the second approach, a method for the determination of the equilibrium density distribution of the mobile ions has been described using experimentally measured values of its flat band voltage under different conditions namely before contamination/activation, after contamination/activation and then after ion drift due to thermal electric stress [12]. This is achieved by deriving an expression for the total ion concentration in terms of the flat band voltages under different cases of assumed distribution such as rectangular, exponential and Gaussian. The computed values of the flat band voltage shifts and the total ion concentration under different assumed distributions are then compared with experimental values. The computed results show that the Gaussian distribution agrees better with experimental results as compared to the rectangular and exponential distributions. This is further supported by an additional computation using the analytical model [13] (see Chap. 7) developed by the authors, which also shows a distribution closer to Gaussian.

Moreover, another work gives a more detailed approach by taking into consideration the influence of all the internal and external electric field on the ions. It studies the mobile ion distribution using the known device parameters and physical constants. The obtained curves (see Figs. 6.9 and 6.10) are in exact conformity with the previous published results. Equilibrium distribution $N(x, t)$, $\zeta(t)$ obtained can be used to study the non-equilibrium processes caused by ion drift under thermal and electric fields in the MOS structures. It can be noticed that by increasing the bias voltage as well as temperature to more positive value, the curve is shifted faster towards the Si–SiO$_2$ interface. The effective voltage in the oxide may be quite different from that of the applied voltage due to redistribution of ions, because Boltzmann law governs a certain part of the energy of the mobile ion.

The subject of mobile ion distribution in oxides of MOS structures, therefore, still remains open for further exploration. It would be worthwhile if a more detailed and deeper analytical approach be undertaken by considering the influence of all the internal and external forces on the ions separately and then combining them suitably to obtain a final equation which is capable of giving the mobile ion distribution in terms of the known device parameters and physical constants. Such an analytical attempt is described in Chap. 7.

References

1. Chou, N.J.: Application of triangular voltage sweep method to mobile charge studies in MOS structures. J. Electrochem. Soc. **118**, 601–609 (1971)
2. Przewlocki, H.M., Marciniak, W.: The triangular voltage sweep method as a tool in studies of mobile charge in MOS structures. Phys. Stat. Sol. A **29**, 265–274 (1975)
3. Tangena, A.G., Middelhoek, J., DeRooij, N.F.: Influence of positive ions on the current–voltage characteristics of MOS structures. J. Appl. Phys. **49**, 2876–2879 (1978)
4. Romanov, V.P., Chaplygin, Yu.A.: Stationary distribution of mobile charge in the dielectric of MOS structures. Phy. Stat. A **53**, 493–498 (1979)
5. Derbenwick, G.F.: Mobile ions in SiO$_2$: potassium. J. Appl. Phys. **48**, 1127–1130 (1977)

6. Snow, E.H., Grove, A.S., Deal, B.E., Sah, C.T.: Ion transport phenomena in insulating films. J. Appl. Phys. **36**, 1664–1673 (1965)
7. Hillen, M.W., Verwey, J.F.: Mobile ions in SiO_2 layers on Si. In: Barbottain, G., Vapaille, A. (eds.) Instabilities in Silicon Devices. North-Holland, Amsterdam (1986)
8. Yon, E., Ko, W.H., Kuper, A.B., IEEE Trans. Elect. Dev. **ED-13**, 276 (1966)
9. Raychaudhuri, A., Ashok, A., Kar, S.: Ion-dosage dependent room-temperature hysteresis in MOS structures with thin oxides. IEEE Trans. Elect. Dev. **38**, 316–322 (1991)
10. Grove, A.S., Deal, B.E., Snow, E.H., Sah, C.T.: Investigation of thermally oxidized silicon surface using MOS structures. Solid Stat. Elec. 145–163 (1965)
11. Bentarzi, H., Bouderbala, R., Mitra, V.: Mobile Ion Distribution in the Oxide of MOS Structure, vol. 2, pp. 85–92. AMSE Press, Malta (1993)
12. Bentarzi, H., Bouderbala, R., Mitra, V.: Determination of the distribution of mobile charges in the oxide of the MOS structure, vol. II, pp. 106–111. ESD'94, Brno (1994)
13. Mitra, V., Bentarzi, H., Bouderbala, R., et al.: A theoretical model for the density-distribution of mobile ions in the oxide of the metal–oxide-semiconductor structures. J. Appl. Phys. **73**, 4287–4291 (1993)
14. Grove, A.S.: Physics and Technology of Semiconductor Devices. Wiley, New York (1967)
15. Nicollian, E.H., Brews, J.R.: MOS Physics and Technology. Wiley, New York (1982)
16. Snow, E.H., Grove, A.S., Deal, B.E., et al.: Ion transport phenomena in insulating films. J. Appl. Phys. **36**, 1664–1673 (1965)
17. Constantinides, A.: Applied Numerical Methods with Personal Computers. McGrawHill, New York (1987)
18. Bentarzi, H., Bouderbala, R., Zerguerras, A.: Simulation of ion density distribution in the gate oxide of MOS structures. AMSE Trans. Modell. Meas. Control **76**, 13–23 (2003)

Chapter 7
Theoretical Model of Mobile Ions Distribution and Ionic Current in the MOS Oxide

7.1 Introduction

It is well known that the presence of the mobile ions in the oxide of a MOS device can greatly influence its electrical characteristics. A number of attempts [1–5] have been made for the experimental determination of the density-distribution of the mobile ions in the oxide. Generally, an intentionally contaminated sample is used for this purpose to overshadow the effect of other types of charge in the oxide. The determination of the density-distribution of mobile ions in such a sample is then carried out by etching its top-layer in steps and each time measuring the total mobile charge in the remaining oxide by using either of the commonly used techniques [1].

A fast, simple and very sensitive technique is triangular voltage sweep (TVS) method that is capable of detecting up to 10^9 mobile ions/cm^2 and is based on the measurement of the displacement current response to a slow linear ramp voltage at elevated temperatures. This yields an ionic displacement current peak whose area is proportional to the total mobile ionic charge. This method has been independently developed by Yamin [3] and Chou [4] who have tested and confirmed its validity using the simpler and electrochemically symmetrical Si(poly)–SiO$_2$–Si(100) MOS structure. This technique is expected to be very useful for routine process and quality control applications. Further, it has been used to study positive mobile charge behavior in the oxide [6, 7].

It is very difficult to obtain a theoretical ionic displacement current using the previous theoretical models of equilibrium distribution of mobile ions [4–6]. However, in the present work, through the use of theoretical model of density distribution of ions developed by authors [8] the ionic displacement I–V characteristic can be easily obtained. The used model is based on the concept that at any point in the oxide, the equilibrium concentration of the mobile ions is attained when the combined mobilizing forces, arising from different origins, become just sufficient to provide necessary activation energy to the ions to surmount the

H. Bentarzi, *Transport in Metal-Oxide-Semiconductor Structures*,
Engineering Materials, DOI: 10.1007/978-3-642-16304-3_7,
© Springer-Verlag Berlin Heidelberg 2011

effective potential well. The resulting theoretical I–V characteristic has been compared with experimental curve using TVS technique. A good agreement between the theoretical and experimental results provides a support to the theoretical model of distribution of ions used here and establishes its merits and usefulness to obtain theoretical ionic current in comparison to other similar studies made in this connection [7–9].

7.2 Theoretical Model of Mobile Ions Density Distribution

Apart from such experimental studies, a few theoretical attempts as discussed in Chap. 6 have been developed to obtain the density-distribution of the mobile ions in the oxide of MOS structures based on the argument that these ions must attain an equilibrium density-distribution under the influence of various internal forces which are operative during the normal course of fabrication of the MOS devices. This work, is another attempt in the same direction, presents a one-dimensional analytical model of the distribution of the mobile ions along the oxide-thickness of a MOS structure. It is based on the concept that the equilibrium concentration of the mobile ions is attained in the oxide when all the mobilizing forces together, arising from different origins, namely, thermal diffusion, internal and external electric fields, become just sufficient to impart the necessary activation energy to the ions to surmount the effective potential well. Unlike the previous studies [4, 6, 10], all these forces acting on a single ion have been obtained here independent of each other from fundamental considerations and finally combined to get the density-distribution of the ions under the equilibrium condition. The results of the present analytical model are consistent and in good agreement with the earlier results.

7.2.1 Preliminary Considerations

Earlier investigations as discussed in Chap. 6 on the transport of the mobile charge in the oxide have led to the conclusion that, in order to produce ionic current, these ions have to overcome potential wells at the interfaces as well as in the bulk of the oxide. The potential wells or ion traps at the interfaces have been found to be somewhat deeper than those existing in the bulk of the oxide. Different estimates about the depth of the potential wells yield values ranging from 0.8 to 0.9 eV near the interfaces and from 0.62 to 0.7 eV in the bulk of the oxide [1, 11–13]. Whereas a little is known [12] about the origin of the ion traps at the interfaces, the ion traps in the bulk of the oxide are supposed to be associated with the periodic inter-atomic potential fluctuations. It will be assumed here that the ion traps at the oxide-interfaces are basically of the same nature as that of the oxide-bulk and all of them have an energy E_{d}. However, at any point near the interfaces, the energy of these ion traps is effectively increased by an amount $q\phi$ (see Fig. 7.1) because

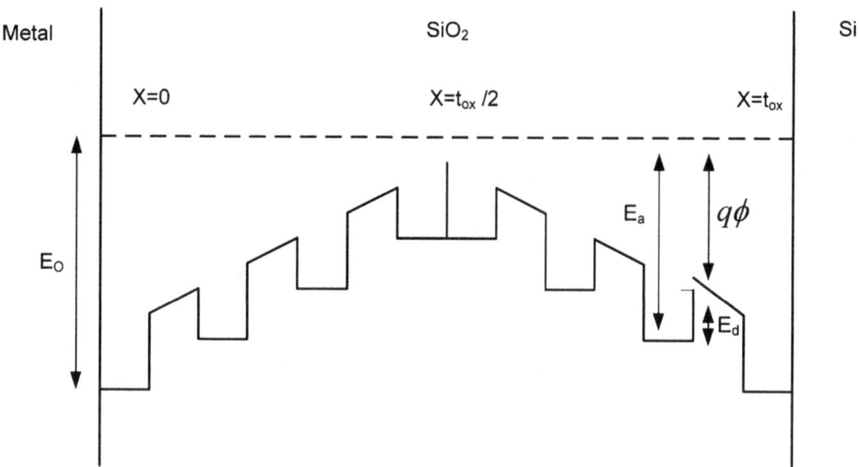

Fig. 7.1 Buildup of the effective activation energy E within the oxide such as a sum of trap-energy E$_d$ and a barrier-energy $q\varphi$

of the creation of a potential barrier by the positive trapped charges of the oxide. In fact, oxide trapped charges are known to be associated with the defects in SiO_2. These oxide traps may have any kind of charge, positive or negative, depending upon the predominant process that is responsible for the flow of the carriers [14] during the oxide formation or subsequent processing. Reference to a number of phenomena is made [15–17] in connection with the conduction-processes in the oxide which may be held responsible for the electron flow through the oxide. The one that may be cited here is the Schottky emission, which consists of thermionic emission across the metal-oxide or oxide-semiconductor interface. In view of the proposed formation of a potential barrier and hence an effective increase of the trap-energy at the interface, the activation energy E_a of the mobile ions may be supposed to vary from a certain maximum energy E_o at the interfaces (metal-oxide interface located at $x = 0$ and oxide-semiconductor interface located at $x = t_{ox}$) to a minimum energy E_d at a certain point x = w close to the middle of the oxide thickness. In fact, the exact location of the point of minimum activation energy E_d, which is the same as point of zero potential barrier, would depend upon the activation energies E_{o1} and E_{o2} at the two interfaces respectively. In the absence of any precise estimate of these traps energies E_{o1} and E_{o2}, it is assumed here that $E_{o1} = E_{o2} = E_o$. Under this condition, the point of minimum energy lies at $x = w$. In the case when the mobile ions are introduced from the metal-oxide interface, all these ions may be assumed to be concentrated in the beginning at $x = 0$. This excessive concentration of positive ions at $x = 0$ gives rise to two forces, one F_1 due to the internal electric field and the other F_2 due to thermal diffusion. These two forces F_1 and F_2 together are much greater at the beginning to overcome the potential barrier existing at the metal-oxide interface. So they mobilize such ions which possess enough energy to surmount the trap-energy E_d. Due to Boltzmann distribution of thermal energy, a certain number of the ions is always supposed to have enough thermal energy to overcome E_d and must respond to the influence of the force $F_1 + F_2$. This should lead, although slowly, to the attainment of equilibrium density distribution of the mobile ions even at room temperature. However, at an elevated temperature, certain more number of the ions would possess the required thermal energy and the attainment of the equilibrium is much faster. During the above process of the attainment of equilibrium, some positive ions drift deeper into the oxide under the action of forces F_1 and F_2 and this process continues till the sum of the forces F_1 and F_2 goes down to a value which is just enough to overcome the local barrier ϕ. Therefore, this marks the condition of equilibrium density distribution of the positive ions. The present model necessitates the evaluation of the forces F_1, F_2 and the potential barrier ϕ, which is performed in the following section.

7.2.2 One-Dimensional Distribution Model of Mobile Ions

As proposed earlier, in the present study, only one-dimensional distribution of the mobile ions is considered. In order to visualize such a model, reference may be made to Fig. 7.2 in which the ion concentration $N(x)$ varies along the oxide-thickness in the direction x so that the whole oxide volume may be considered to be composed of thin successive parallel layers of positive ions like A, B and C. Let the layers A, B and C be located at $x_o - \lambda$, x_o and $x_o + \lambda$ and have ion-concentration $N_{x_o-\lambda}$, N_{x_o} and $N_{x_o+\lambda}$ respectively such that the mutual distance λ between the adjacent layers does not vary significantly within a small thickness element at x. At any point P of the layer B, the electric field $\zeta_{x_o-\lambda}$ due to all the positive ions situated in the layer A should be pointing towards right and can be written as:

$$\zeta_{x_o-\lambda} = \frac{\sigma_{x_o-\lambda}}{2\varepsilon_0\varepsilon_{ox}} \tag{7.1}$$

where $\sigma_{x_o-\lambda}$ is the charge density in the layer A. Similarly, the electric field $\zeta_{x_o+\lambda}$ on the same point P due to all the positive ions of layer C should be pointing towards left and can be written as:

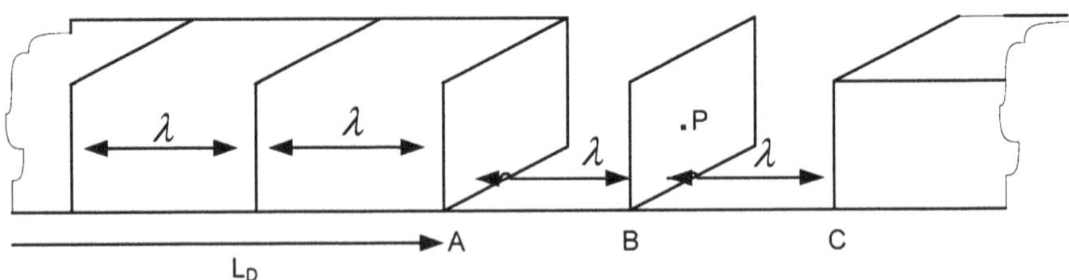

Fig. 7.2 Cross-sectional view of the oxide thickness showing parallel mono-molecular layers as A, B, and C with inter-distance λ

$$\zeta_{x_o+\lambda} = \frac{\sigma_{x_o+\lambda}}{2\varepsilon_o\varepsilon_{ox}} \tag{7.2}$$

The forces $F_{x_o-\lambda}$ and $F_{x_o+\lambda}$ on a single positive ion of charge q located at P due to the electric field $\zeta_{x_o-\lambda}$ and $\zeta_{x_o+\lambda}$ respectively can be written as:

$$\zeta_{x_o-\lambda} = \frac{q\sum\sigma_{x_o-\lambda}}{2\varepsilon_o\varepsilon_{ox}} \tag{7.3}$$

and

$$F_{x_o+\lambda} = \frac{q\sum\sigma_{x_o+\lambda}}{2\varepsilon_o\varepsilon_{ox}}. \tag{7.4}$$

where the summation is to be carried out on all such layers on the right side of layer A in Eq. 7.3 and on the left side of layer C in Eq. 7.4 which can have influence on the layer B. The net force F_1 experienced by a single ion at P comes out to be

$$F_1 = q\frac{\left(\sum\sigma_{x_o-\lambda} - \sum\sigma_{x_o+\lambda}\right)}{2\varepsilon_o\varepsilon_{ox}}. \tag{7.5}$$

Let L_D be the Debye length at the layer B, that is, the distance up to which the positive ions on the left or the right of layer B can exert influence on the considered single ion at P. Considering the whole region up to a distance L_D on the right side of the layer B to be divided into similar parallel layers of positive ions with an inter-distance λ, the right hand term of Eq. 7.3 can be easily evaluated since it simply equals to the sum of the following arithmetic series containing (L_D/λ) terms:

$$\frac{q}{2\varepsilon_o\varepsilon_{ox}}\sum\sigma_{xo-\lambda} = \frac{q^2}{2\varepsilon_o\varepsilon_{ox}}\left[\lambda N_{xa} + \lambda\left(N_{xo} - \lambda\frac{\delta N_{xo}}{\delta x}\right) + \gamma\left(N_{xo} - 2\lambda\frac{\delta N_{xo}}{\delta x}\right) + \ldots\right]$$

or

$$\frac{q}{2\varepsilon_o\varepsilon_{ox}}\sum\sigma_{xo-\lambda} = \frac{q^2}{2\varepsilon_o\varepsilon_{ox}}\left[\lambda\left(N_{x\grave{a}}\frac{L_D}{\lambda} - \lambda\frac{\delta N_{xo}}{\delta x}\frac{L_D}{2\lambda}\right)\right] \tag{7.6}$$

Similarly right-hand side of Eq. 7.4 can be written as

$$\frac{q}{2\varepsilon_o\varepsilon_{ox}}\sum\sigma_{xo+\lambda} = \frac{q^2}{2\varepsilon_o\varepsilon_{ox}}\left[\lambda\left(N_{x\grave{a}}\frac{L_D}{\lambda} + \lambda\frac{\delta N_{xo}}{\delta x}\frac{L_D}{2\lambda}\right)\right] \tag{7.7}$$

Substitution of the expressions of $\sum\sigma_{x_o-\lambda}$ and $\sum\sigma_{x_o+\lambda}$ in Eq. 7.5 gives

$$F_1 = -\frac{q}{2\varepsilon_o\varepsilon_{ox}}\frac{\delta N_{xo}}{\delta x}L_D^2 \tag{7.8}$$

It may be pointed out that the force F_1 is directing down the concentration gradient in the case of positive ions. Yet another force F_2 is supposed to come into play on the considered single ion at P in the same direction that is due to thermal diffusion because of the concentration gradient $\frac{\delta N_{xo}}{\delta x}$. This force can be obtained from kinetic considerations. In case all these positive ions are free to move and have only thermal energy, they can be treated like free molecules of a gas with concentration gradient $\frac{\delta N_{xo}}{\delta x}$. The difference of the partial pressure of this so-called gas on the two faces of a volume-element between the layers A and C is a measure of the collective force $\sum F_2$, which is acting on all the positive ions contained in the considered volume-element with unit area of cross-section. In this way, the average force F_2 can be written as

$$F_2 = \frac{\frac{1}{3}mC^2 N_{xo-\lambda} - \frac{1}{3}mC^2 N_{xo+\lambda}}{N_{xo}(2\lambda)}, \tag{7.9}$$

or

$$F_2 = -\frac{C^2}{3N_{xo}}\frac{\delta N_{xo}}{\delta x}. \tag{7.10}$$

where m is the mass of ion and C the R.M.S velocity. Considering the positive ions to be in thermal equilibrium with the bulk of the oxide layer at Tk,

$$\frac{1}{2}mC^2 = \frac{3}{2}kT, \tag{7.11}$$

where k is the Boltzman constant. Equation 7.10 combined with Eq. 7.11 gives

$$F_2 = -\frac{kT}{N_{xo}}\frac{dN_{xo}}{dx}. \tag{7.12}$$

The forces F_1 and F_2 are important for establishing the equilibrium-distribution of the mobile ions soon after these ions are introduced in the oxide. Since these ions require enough activation energy to move from one place to another, the process of attaining this equilibrium distribution may not be as fast as it would be if the ions were completely free. As already discussed, the total activation energy E_a is made up of two parts. The first is needed to overcome the trap-energy E_d that is constant throughout the width of the oxide-layer and the other part is required to overcome the barrier energy $q\phi$ that varies with the position of the ion. Whereas the first part E_d is provided by the thermal energy of the ion, the other part $q\phi$ is built-up by the mobilizing forces F_1 and F_2. Thus the speed, how quickly the equilibrium density-distribution can be obtained, depends upon the thermal energy and hence the temperature of the oxide. Obviously, at higher temperature the attainment of equilibrium density-distribution is faster. In any case, the equilibrium distribution of ions occurs when the sum of the forces F_1 and F_2 comes down to such a limiting value F_3 which is required to overcome the barrier energy $q\phi$ so that

$$F_1 + F_2 = F_3. \tag{7.13}$$

In order to evaluate the force F_3 which is required to overcome the potential barrier ϕ, it is essential to know how the barrier ϕ varies with the depth of the oxide as measured from the metal-oxide interface at $x = 0$. This can be known under a simple assumption that the positively trapped charges, which are responsible for the creation of the potential barrier ϕ, have uniform density N_t along the thickness of the oxide. The potential ϕ at any point distant x from the metal-oxide interface, therefore, can be easily obtained by solving the Poisson's equation under proper boundary conditions. In the light of the earlier discussions, these boundary-conditions may be fixed such that the electric potential ϕ as well as the electric field $E = -\frac{d\phi}{dx}$ both are zero at $x = t_{ox}/2$. Although such a boundary condition does not reconcile with the idea of an uniform distribution of oxide trapped charges continuously from one end of oxide to another, but it can do so under a limitation when the point of zero potential and field does not exactly lie at $x = t_{ox}/2$ but little away from it separately for the two halves of the oxide. Under this physical picture, the boundary condition can be fixed separately for the two halves such that $\phi = 0$ and $-\frac{d\phi}{dx} = 0$ at x_1 for the first half and at x_2 for the second where x_1 is a little smaller than $t_{ox}/2$ and x_2 a little greater than $t_{ox}/2$. The expressions for the electric potential (which is related to activation of the ion) and that of the electric field (which is related to the activation force) can be obtained separately for the two halves under the above the boundary conditions. Any set of two such expressions, obtained in this way for the two halves, can be then combined into common expression which may be applicable for the whole oxide-thickness under the limit when $x_1 = x_2 \approx \frac{t_{ox}}{2}$. Proceeding at first for the first half under the boundary conditions $\phi = 0$, $-\frac{d\phi}{dx} = 0$ at $x = x_1$, the potential ϕ_1 and the field ζ_1 can be written respectively as

$$\phi_1 = \frac{-qN_t}{2\varepsilon_o\varepsilon_{ox}}\left(\frac{t_{ox}}{2} - x\right)^2 \tag{7.14}$$

and

$$\zeta_1 = \frac{qN_t}{2\varepsilon_o\varepsilon_{ox}}\left(\frac{t_{ox}}{2} - x\right) \tag{7.15}$$

Similarly proceeding for the second half under the boundary conditions $\phi = 0$ and $-\frac{d\phi}{dx} = 0$ at $x = x_2$ with respect to the origin at $x = x_2$, the potential ϕ_2' and the field ζ_2' can be written respectively as

$$\phi_2' = \frac{-qN_t x^2}{2\varepsilon_o\varepsilon_{ox}} \tag{7.16}$$

and,

$$\zeta_2' = \frac{qN_t x}{2\varepsilon_o\varepsilon_{ox}} \tag{7.17}$$

After transforming the coordinates with respect to the origin from $x = x_2$ to $x = 0$, the new expressions for the potential ϕ_2 and the electric field ζ_2 are

$$\phi_2 = \frac{-qN_t}{2\varepsilon_o\varepsilon_{ox}}\left(\frac{t_{ox}}{2} - x\right)^2 \tag{7.18}$$

and

$$\zeta_2 = \frac{qN_t}{2\varepsilon_o\varepsilon_{ox}}\left(\frac{t_{ox}}{2} - x\right) \tag{7.19}$$

Since Eqs. 7.18 and 7.19 merge into Eqs. 7.14 and 7.15 respectively in the limiting case when $x_1 = x_2 = \sim t_{ox}/2$, the following common equations for the potential ϕ and the electric field ζ can be written which are applicable for the whole oxide-thickness:

$$\phi = \frac{-qN_t}{2\varepsilon_o\varepsilon_{ox}}\left(\frac{t_{ox}}{2} - x\right)^2 \tag{7.20}$$

and

$$\zeta = \frac{qN_t}{2\varepsilon_o\varepsilon_{ox}}\left(\frac{t_{ox}}{2} - x\right) \tag{7.21}$$

The difference between the activation energy E_a and the trap-energy E_d gives at any point x the value of $q\phi$ so that

$$E_a - E_d = \frac{-q^2 N_t}{2\varepsilon_o\varepsilon_{ox}}\left(\frac{t_{ox}}{2} - x\right)^2 \tag{7.22}$$

From Eq. 7.22, the difference $E_o - E_d$ between the maximum and minimum activation energy can be written as

$$E_o - E_d = \frac{-q^2 N_t t_{ox}^2}{8\varepsilon_o\varepsilon_{ox}} \tag{7.23}$$

Equation 7.22 combined with Eq. 7.23, yields the following expression for the activation energy $E_a(x)$:

$$E_a(x) = (E_o - E_d)\left(1 - \frac{2x}{t_{ox}}\right)^2 + E_d \tag{7.24}$$

The activation force F_3, which is required to provide this much activation energy E_a, can be written as

$$F_3 = -\frac{dE_a(x)}{dx} = \frac{4(E_o - E_d)}{t_{ox}^2}(t_{ox} - 2x) \tag{7.25}$$

Finally, Eq. 7.13 combined with Eqs. 7.8, 7.12 and 7.25, gives the following condition for the equilibrium distribution of the mobile ions in the oxide:

$$-\frac{q^2 L_{\mathrm{D}}^2}{2\varepsilon_o \varepsilon_{ox}}\frac{dN(x)}{dx} - \frac{kT}{N(x)}\frac{dN(x)}{dx} = -\frac{dE_a(x)}{dx} = \frac{4(E_o - E_d)}{t_{ox}^2}(t_{ox} - 2x). \qquad (7.26)$$

It may be pointed out here that Eq. 7.26 does not include the effect of the applied voltage V_A and that of any charge which may exist in the oxide other than the mobile positive ions and positively charged such oxide-traps. However, in practice two more types of charge are supposed to exist in the oxide, namely, the fixed oxide charge and interface trapped charge both of which are located at the Si–SiO$_2$ interface.

In most cases, these two types of charge amount to an effective negative gate-voltage $-V_F$. The mobilizing force F_4 due to the combined effect of the voltages V_A and V_F on an ion can be written as:

$$F_4 = \frac{q(V_A - V_F)}{t_{ox}} \qquad (7.27)$$

After taking into account this force F_4, Eq. 7.26 becomes

$$\frac{q(V_A - V_F)}{t_{ox}} - \frac{q^2 L_{\mathrm{D}}^2}{2\varepsilon_o \varepsilon_{ox}}\frac{dN(x)}{dx} - \frac{kT}{N(x)}\frac{dN(x)}{dx} = \frac{4(E_o - E_d)}{t_{ox}^2}(t_{ox} - 2x). \qquad (7.28)$$

In an assembly of distributed point-charges under equilibrium, the Debye length L_{D} is the distance over which the effect of any single point-charge dies out and is expressed [18] in terms of the equilibrium-density N_{xo} of such point-charges. This expression of L_{D}, as applied to the case of mobile oxide-charges, can be written as:

$$L_{\mathrm{D}} = \sqrt{\frac{\varepsilon_o \varepsilon_{ox} kT}{q^2 N_{xo}}}. \qquad (7.29)$$

Using the expression of Debye length in Eq. 7.28 and then integrating the resulting equation under the boundary-condition $N(x) = No$ at $x = 0$, the following relation for the equilibrium density-distribution of mobile ions can be obtained:

$$Ln\frac{N(x)}{N_o} = \frac{8(E_o - E_d)}{3kTt_{ox}^2}(x^2 - xt_{ox}) + \frac{2}{3}\frac{q(V_A - V_F)x}{kTt_{ox}}, \qquad (7.30)$$

or

$$N(x, V) = N_O \cdot \exp\big(A.(x^2 - xt_{ox}) + B(V_A - V_F)x\big), \qquad (7.31)$$

where, $A = \frac{8(E_o - E_d)}{3kTt_{ox}^2}$ and $B = \frac{2q}{3kTt_{ox}}$, E_d the trap energy, E_o the activation energy at the interface, N_o is the assumed total charges per unit area and V_A the applied voltage.

Equation 7.31 can be used to obtain the concentration-profile (N-x curve) of the mobile ions along the oxide-thickness. However, experimental profiles are so carried out that they yield the total mobile ions density $(N_{xo})_{tot}$, which is the average ion-density in the oxide from x to t_{ox}, as a function of x. In order to compare the results of the present theoretical model with the experiments, it would be therefore worthwhile to obtain an expression for $(N_{xo})_{tot}$ which can be easily done with the help of Eq. 7.31. The following expression for $(N_{xo})_{tot}$ can be written:

$$(N_{xo})_{tot} = \frac{1}{(t_{ox} - x_0)} \int_0^{t_{ox}} N(x)dx \tag{7.32}$$

Further, the flat-band-voltage shift ΔV_{FB} due to the introduction of mobile charges, distributed according to the $N(x)$ profile given by Eq. 7.31, in the oxide can be written as [1]

$$\Delta V_{FB} = \frac{q}{\varepsilon_{ox}\varepsilon_o} \int_0^{t_{ox}} xN(x)dx \tag{7.33}$$

7.3 I–V Characteristic Determination

The foregoing model can also be used to calculate the influence of the oxide ion transport on the I–V characteristics of MOS structures [8]. It is assumed that if the change of applied voltage across the oxide is sufficiently slow (quasi-static), the instantaneous ion distribution can always be approximated by the equilibrium distribution. Under the assumption that the space charge density in the oxide is wholly made up of mobile ions and the applied voltage V_A across it varies linearly with time t, the displacement current flowing through the metal gate can be written as

$$i(V) = \pm \alpha C_{LF}(V) \pm \alpha q \frac{d}{dV} \int_0^{t_{ox}} \frac{x}{t_{ox}} N(x, V)dx \tag{7.34}$$

where $C_{LF}(V)$ is the low frequency MOS capacitor, $N(x, V)$ the mobile ions distribution per unit area. The theoretical model [8] for the density distribution of ions in the oxide, which applies fairly well to the quasi-static state, can be used after setting $z = \frac{x}{t_{ox}}$ and $dz = \frac{dx}{t_{ox}}$, then Eq. 7.31 may be reduced to

$$N(z, V) = N_o \exp\left(A'(z^2 - z) + B'(V_A + V_F)z\right) \tag{7.35}$$

where, $A' = A.t_{ox}^2$ and $B' = B.t_{ox}$.

It may be pointed out that V_F includes the effect of work function difference as well as other types of charge, which are located at Si–SiO$_2$ interface. This quasi-static equilibrium is supposed to occur at elevated temperatures, in the range above 150°C. Consequently the low frequency MOS capacitance will be equal to oxide capacitance per unit area, $C_{LF}(V) = C_{ox}$. Under this condition, substitution of relation Eq. (7.35) in Eq. (7.34), gives [7]

$$i(V) = \pm\alpha.C_{ox} \mp \alpha.q.N_oB'.t_{ox}.\int_0^1 z^2 \exp\left(A'(z^2 - z) + B'(V_A + V_F)z\right)dz \quad (7.36)$$

Equation 7.36 can be used for numerical computation of the I–V characteristic. Since the MOS behavior is symmetric about $V = 0$, the above computation can be carried out by making use of only one polarity of the voltage [9].

7.4 Experimental Results and Discussion

Values of the minimum and maximum activation energy as required in the computation are taken to be $E_o = 0.8$ to 0.9 eV [1, 11] and $E_d = 0.62$ to 0.70 eV [12, 13]. Based upon the above experimental values of E_o and E_d, the maximum and minimum limits of the quantity $(E_o - E_d)$ may be fixed to be nearly 0.3 and 0.1 eV respectively. Therefore, in the present computation, three different values of $(E_o - E_d)$ equal to 0.1, 0.2 and 0.3 eV are used. Theoretical plots for the mobile ion density-distribution, as obtained with the help of Eq. 7.31, are given in Fig. 7.3 for $(E_o - E_d) = 0.1$, 0.2 and 0.3 eV respectively at $T = 300$ K and using $N_o = 2 \times 10^{18}$ cm^{-3}, oxide-thickness 540 nm, and $(V_A - V_F) = -0.35$ V. The same curves have been repeated for a number of values of applied effective voltage $(V_A - V_F)$ ranging from -0.5 to $+0.5$ V. These results are reproduced in Fig. 7.4 for $(E_o - E_d) = 0.3$ eV. Finally, the effect of the temperature is studied by repeating a few typical curves at different values of the temperature, which are reproduced in

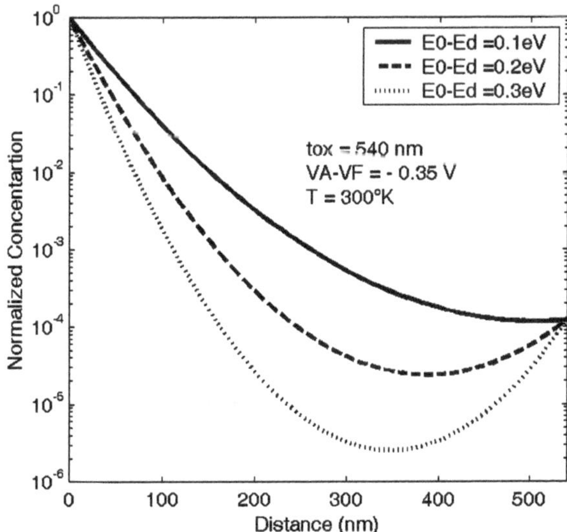

Fig. 7.3 Effect of the parameter $E_0 - E_d$ on the theoretical concentration profiles of the mobile ions with different values **a** 0.1 eV, **b** 0.2 eV and **c** 0.3 eV

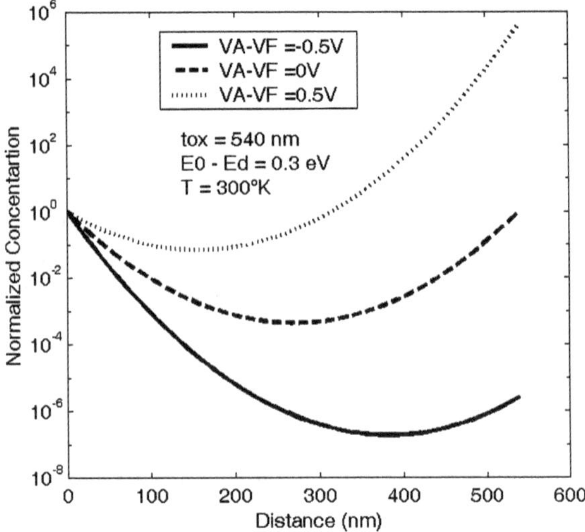

Fig. 7.4 Effect of the effective gate voltage $V_A - V_F$ on the theoretical concentration profiles of the mobile ions using different values **a** −0.5 V, **b** 0 V and **c** 0.5 V, for $E_0 - E_d = 0.3$ eV

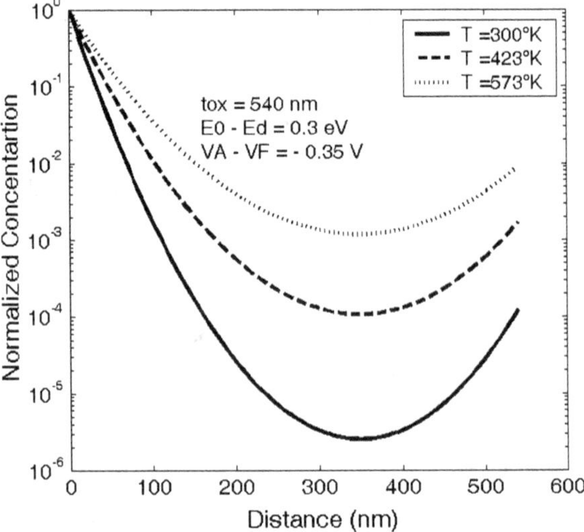

Fig. 7.5 Effect of the temperature T on the theoretical concentration profiles of the mobile ions using different values **a** 300 K, **b** 423 K and **c** 573 K, for $E_0 - E_d = 0.3$ eV

Figs. 7.5, and 7.6. It may be seen that before the effective gate-voltage $(V_A - V_F)$ acquires enough positive value, the theoretical curves for the density-distribution of the mobile ions are nearly exponential in nature and the ion-density falls rapidly as the distance is moved from the metal-oxide interface. However, as the effective gate-voltage $(V_A - V_F)$ becomes more and more positive, the concentration of the ions starts building up at the SiO_2–Si interface and the distribution-curve takes a U-shape as shown in Figs. 7.7 and 7.8. Finally, when $(V_A - V_F)$ attains enough positive value, the distribution-curve shows an exponential increase of the ions towards the

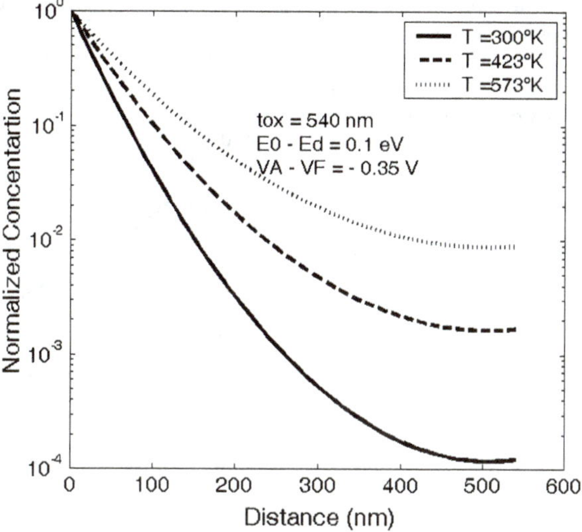

Fig. 7.6 Effect of the temperature T on the theoretical concentration profiles of the mobile ions using different values **a** 300 K, **b** 423 K and **c** 573 K, for $E_0 - E_d = 0.1$ eV

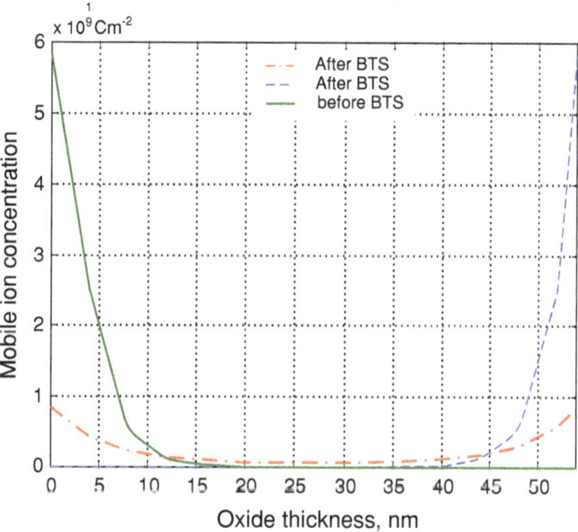

Fig. 7.7 Theoretical curves of mobile ions density distribution at T = 300 K with $No_{tot} = 2 \times 10^{18}$ cm^{-3} and oxide thickness = 54 nm

Si–SiO$_2$ interface. These curves are in exact conformity with the previous results [11, 19, 20] as illustrated in Fig. 7.9.

To further test the validity of the present analytical model, the flat-band-voltage shift ΔV_{FB} has been calculated with the help of Eq. 7.32 and compared with the experimentally obtained value. For example, ΔV_{FB} can be obtained from the experimentally determined C–V curves before and after intentional contamination of the device. Such C–V curves are reproduced from two earlier measurements [2, 21] in Figs. 7.10 and 7.11 respectively. The values of ΔV_{FB}, as obtained from each of these figures, are compared with the corresponding theoretical values in

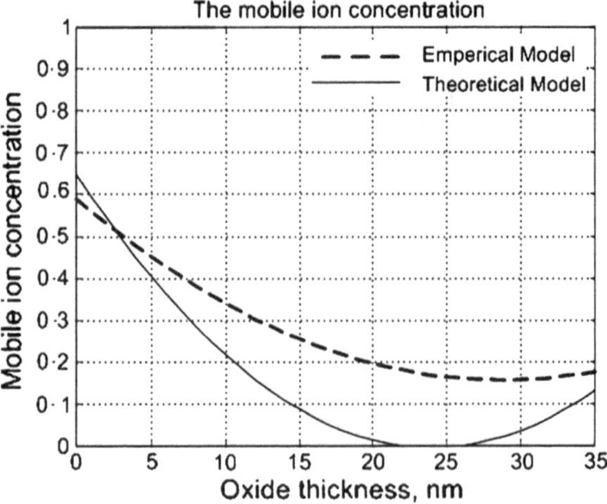

Fig. 7.8 Comparison of the empirical and theoretical model of concentration profile of mobile ions within the oxide of 35 nm thickness

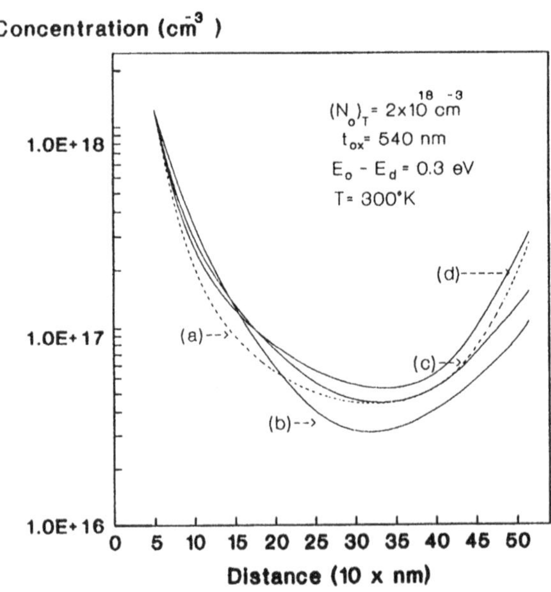

Fig. 7.9 Theoretical and experimental curves of total concentration profiles of mobile ions in MOS structure: **a** experimental curve, theoretical curves for different values of $V_A - V_F$: **b** −0.2 V, **c** −0.185 V and **d** −0.17 V

Table 7.1. The two sets of the value of ΔV_{FB} show quite a good agreement. The consistency, with which the theoretical results agree with the experiments, are indicative of the validity of the proposed analytical model of the distribution of the mobile ions in the oxide of the MOS structures. The computed parameter values such as the total density and the flat-band voltage shift have been carried out by numerical methods. The role of temperature needs special discussion. It may be seen that the present study does not put any such constraint that equilibrium density-distribution can not be attained without rise of temperature. In fact, equilibrium density-distribution can be attained even at room temperature

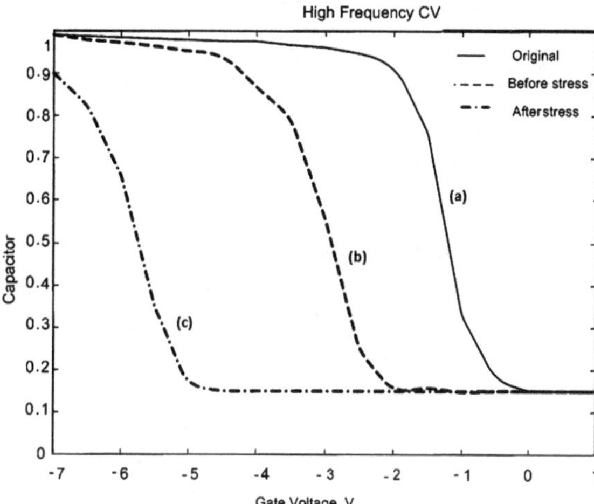

Fig. 7.10 Capacitance–voltage curves of MOS structures. **a** before contamination, **b** after contamination and **c** after 5 min of biasing (Reproduced with permission IEEE [21])

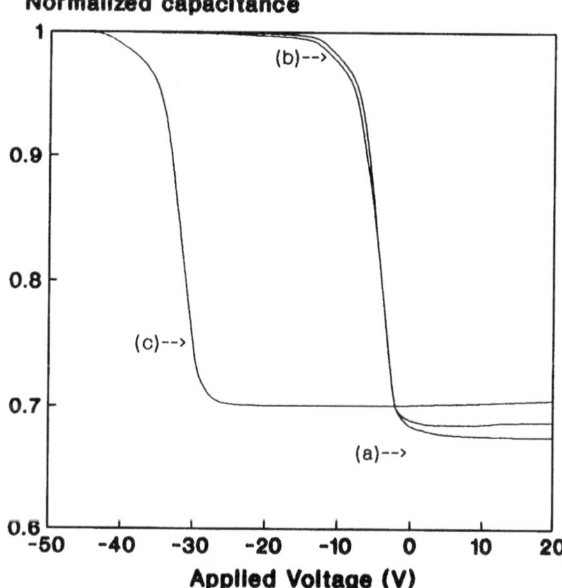

Fig. 7.11 Capacitance–voltage curves of MOS structures: **a** before activation, **b** after activation and **c** after 2700 s of biasing (Reprinted with permission [2])

according to the present model although it may take place quite slowly. It is because the role of temperature is to provide thermal energy to the ions so that they are stimulated to overcome the trap-energy of the potential well, which is nearly 0.66 eV. Assuming the energy of these ions to be governed by Boltzmann law, at least some ions is always supposed to have energy above 0.66 eV and these ions will contribute to the ionic current. Obviously, at the room temperature, the certain number of such ions, which contribute to the ionic current, is very small and the attainment of the equilibrium density-distribution may take quite a long time.

Table 7.1 Comparison between experimental and theoretical values of the flat-band voltage shift resulting from contamination/activation of sodium ions

Ref.	$(N_{xo})_t$ (cm^{-3})	t_{ox} (nm)	$E_a - E_d$ (eV)	$V_A - V_F$ (V)	Expt. ΔV_{FB} (V)	Cal. ΔV_{FB} (V)
[2]	1.5×10^{17}	200	0.1	-0.45	1.4	1.4
[21]	1.2×10^{18}	35	0.1	-0.09	1.6	1.5

At high temperature this process is faster as may be seen from the exponential term in Eq. 7.31 which reveals that at any point x the concentration is more at higher temperature. The present study also envisages that equilibrium density-distribution can be attained under the influence of internal electric field and thermal diffusion alone without the application of any external voltage. However, it may again take a long time to do so. The importance of thermal-electric stress in the experiments concerning ion-transport can be understood from the fact that in the absence of either heat or biasing the rate of drift and hence the resulting ionic current may fall quite below the detection-limits. It may be pointed out further that the role of heat is mainly to stimulate the ions to overcome the trap-energy of the potential well although it plays a role in the ion-drift also to some extent by way of thermal diffusion. Equation 7.24 can give more insight into the role of the trap-energy of the potential wells. In fact the drifting field is obtained by differentiating the activation energy of Eq. 7.24 in which the constant term E_d contributes nothing. However, this energy E_d does put a constraint on the number of ions, which at any time are readily available for the drift by the drifting field. As the height of the potential barrier is greater at points closer to the metal-oxide interface, the effective depth of the potential well is increased at such points reducing thereby the number of ions readily available for the drift. Therefore, the ions, trapped in the potential wells closer to the metal-oxide interface, contribute to less ionic current than those located away from it. Obviously the drift of such ions can be increased by increasing the temperature as per requirement of Eq. 7.31. It is this effect which has led to the earlier conclusion that the ionic current is emission-limited and that it needs at first de-trapping of the ions by heating before they can be drifted by the applied electric field [22]. Finally, this analytical model yields the distribution expression in a closed form identical to Eq. 7.4. The mobile ions distribution curves as obtained by using this analytical model are also given in Fig. 7.7 for comparison with the distribution-curves as obtained by computation using Eq. 6.42 in Chap. 6. A closed agreement between the theoretical and computed distribution provides an experimental support to the present analytical model because the mobile ion-distribution in the oxide of the MOS structures as obtained in Chap. 5 is generally based on the experimentally measured C–V curves.

In practice, expression (7.36) for the displacement current is valid under quasi-static conditions, that is, under the conditions when the voltage is so slow (less than 100 mV/s) that the ionic charge distribution $N(x, V)$ is always under equilibrium with externally applied voltage. To carry out such an experiment under suitable conditions using TVS technique, the general experimental set up system

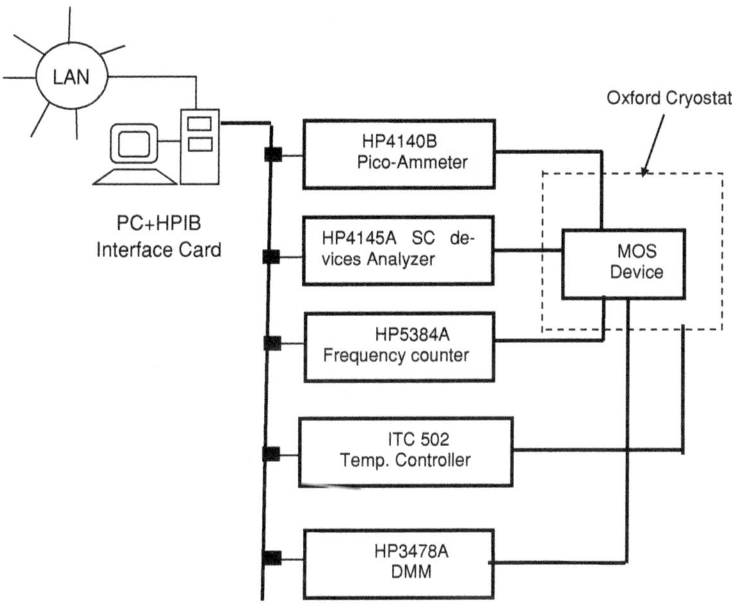

Fig. 7.12 Experimental setup for TVS technique

shown in Fig. 7.12 may be used. A triangular gate voltage sweep, generated by a suitable power supply P, is applied to MOS structure. A programmable dc voltage supply HP 4140 B is used to apply the desired signal on the device mounted within cryostat Oxford. Programmable temperature controller ITC4 is used to keep the experiment temperature within desired limits (intervals). The current flowing in the gate of the device was measured by the HP 4140B Pico-ammeter. Software program has been developed so that PC, with help of IEEE HPIB interface driver, may read and control data from different HP meters and process them to give the desired I–V characteristics. The experimental devices consist of a number of MOS structures, fabricated by ES2 (European Silicon Structure (ESS), Paris, France) laboratory using the 2 μm process, which were made available for the present experimentation by courtesy of CDTA (Center de Development des Technologies Avancées, Algers, Algeria). All the devices from ES2 laboratory have a common value of doping concentration 7.6×10^{15} cm^{-3} of the substrate and an oxide thickness varying from 10 to 40 μm. The validity of the theoretical model has been verified by carrying out experimental measurements on a number of MOS devices. In practice, the condition $V_{ox} = 0$ is obtained at an applied external voltage V_A equal to voltage V_F arising from the work-function difference between metal and silicon and other types of charge at silicon-oxide interface. Therefore, the top of the peak in I–V characteristic does not occur at $V_A = 0$, but at $V_A = -V_F$. The Experimental I–V curve is compared with the computed curve. For computation, N_o is determined by integration of experimental curve [13]. Further, other parameters such as activation energy etc. are taken from reference [12, 13].

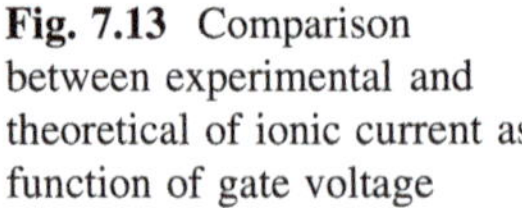

Fig. 7.13 Comparison between experimental and theoretical of ionic current as function of gate voltage

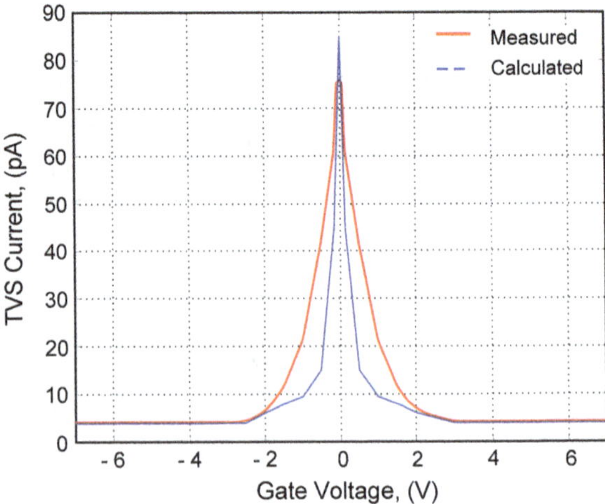

Figures 7.13 and 7.14 show a fairly good agreement between the experimental and computed I–V curves of which the latter is the outcome of the cited theoretical model of mobile ion distribution in MOS Structures.

7.5 Conclusion

Almost all the previous theoretical approaches are based on the direct or indirect use of the ion transport equation that has been obtained in different investigations using different physical ideas. The ion transport equation so obtained is then coupled with the Poisson's equation in all the methods to get a differential equation either in terms of electric field or ion density. These methods however differ in their subsequent analysis of solving the differential equation to obtain the mobile ion density profiles. The differential equation so obtained is either not solved completely as in the first approach or it is solved under certain assumed boundary conditions using assumed value of parameters such as electric potential, field or charge in the oxide. Thus, none of the studies is capable of giving precise distribution for a given device under given temperature-bias condition. Besides, all the methods are incapable of giving the effect of individual factors on the mobile ion distribution. However, our developed model [8] gives a more detailed and deeper analytical approach by considering the influence of all the internal and external forces on the ions separately and then combining them suitably to obtain a final equation. The obtained equation can give the mobile ion distribution in terms of the known device parameters and physical constants. It is very difficult to obtain a theoretical ionic displacement current using the previous theoretical models of equilibrium distribution of mobile ions. However, through the use of this theoretical model of density distribution of ions, the ionic displacement I–V characteristic can be easily obtained. The obtained curves (see Fig. 7.13 and 7.14) are in

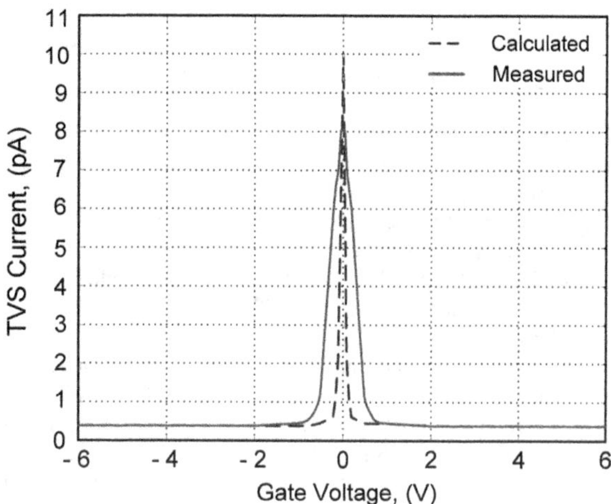

Fig. 7.14 Comparison between experimental and theoretical curves of gate current

exact conformity with the experimental results. The consistency, with which the theoretical results agree with the experiments, is indicative of the validity of the theoretical model of the distribution of the mobile ions in the oxide of the MOS structures.

Finally, the subject of mobile ion distribution in oxides of MOS structures, therefore, remains open for further research exploration.

References

1. Nicollian, E.H., Brews, J.R.: MOS Physics and Technology. Wiley, New York (1982)
2. Grove, A.S., Deal, B.E., Snow, E.H., Sah, C.T.: Investigation of thermally oxidized silicon surface using MOS structures. Solid-St. Electr. 145–163 (1965)
3. Yamin, M.: Charge storage effects in silicon dioxide films. IEEE Trans. Elect. Dev. **ED-12**, 88–96 (1965)
4. Chou, N.J.: Application of triangular voltage sweep method to mobile charge studies in MOS structures. J. Electrochem. Soc. **118**, 601–609 (1971)
5. Hickmott, T.W.: Thermally stimulated ionic conductivity of sodium in thermal. J. Appl. Phys. **46**, 2583–2598 (1975)
6. Przewlocki, H.M., Marciniak, W.: The triangular voltage sweep method as a tool in studies of mobile charge in MOS structures. Phys. Stat. Sol. (a) **29**, 265–274 (1975)
7. Bentarzi, H., Bouderbala, R., Zerguerras, A.: Ionic current in MOS structures. Ann. Telecommun. **59**(3–4), 471–478 (2004)
8. Mitra, V., Bentarzi, H., Bouderbala, R., et al.: A theoretical model for the density distribution of mobile ions in the oxide of the metal-oxide-semiconductor structures. J. Appl. Phys. **73**, 4287–4291 (1993)
9. Kuhn, M., Silversmith, D.J.: Ionic contamination and transport of mobile ions in MOS structures. J. Electrochem. Soc. **118**, 966–970 (1971)
10. Romanov, V.P., Chaplygin, Yu.A.: Stationary distribution of mobile charge in the dielectric of MOS structures. Phy. Stat. (a) **53**, 493–498 (1979)
11. Hillen, M.W.: Dynamic behavior of mobile ions in SiO_2 layers. In: Partelides, S.T. (ed.) The Physics of SiO_2 and Its Interface. Pergamon, New York (1978)

12. Hillen, M.W., Verwey, J.F.: Mobile ions in SiO_2 layers on Si. In: Barbottain, G., Vapaille, A. (eds.) Instabilities in Silicon Devices. Amsterdam (1986)
13. Stagg, J.P.: Drift mobilities of Na+ and K+ ions in SiO_2 films. Appl. Phys. Lett. **31**, 532–533 (1977)
14. Sze, S.M.: Physics of Semiconductor Devices. Wiley, New York (1981)
15. Boudry, M.R., Stagg, J.P.: the kinetic behavior of mobile ions in the $Al–SiO_2–Si$ System. J. Appl. Phys. **50**, 942–950 (1979)
16. O'Dwyer, J.J.: The Theory of Electrical Conduction and Breakdown in Solid Dielectrics. Clarendon, Oxford (1973)
17. Av-Rom, M., Shatzkes, M., Distefano, T.H., Cadoff, I.B.: The nature of electron tunneling in SiO_2. In: Pantelider, S.T. (ed.) The Physics of SiO_2 and its Interface. Pergamon, New York (1978)
18. Grove, A.S.: Physics and Technology of Semiconductor Devices. Wiley, New York (1967)
19. Tangena, A.G., Middelhoek, J., DeRooij, N.F.: Influence of positive ions on the current–voltage characteristics of MOS structures, J. Appl. Phys. **49**, 2876–2879 (1978)
20. Feigl, F.J., Butler, S.R.: Electrochem. Soc. Meeting, Atlanta Oct 9–14 (1977)
21. Raychaudhuri, A., Ashok, A., Kar, S.: Ion-dosage dependent room-temperature hystersis in MOS structures with thin oxides. IEEE Trans. Elect. Dev. **38**, 316–322 (1991)
22. Hofstein, S.R.: IEEE Trans. Elect. Dev. **ED-14**, 749 (1976)

Index